'One of the *best five books*
I have ever had the privilege to read.'
Tim Smit, *The Eden Project*

'Books have been written about entire countries that contain a less interesting cast of characters than Lewis-Stempel's account of one field on the edge of Wales. Foxes, red kites and voles become as *intricately shaded* as characters in an HBO drama . . . Not every English meadow contains such a vast variety of wildlife as Lewis-Stempel's but his *immense, patient powers of observation* – along with a flair for the anthropomorphic – mean he is able to offer a portrait of animal life that's rare in its colour and drama.'
Tom Cox, *Observer*

'[John Lewis-Stempel] has a sharp eye, a fluent pen and that omnivorous, innocently English curiosity about wild creatures . . . There are *lyrical* moments aplenty but this is not the cloying "regardez-moi maman" nature writing. [His] tone is level, involved, *humorous* and even self-deprecating . . . This is a *rich, interesting* book, generously studded with raisins of curious information.'
Angus Clarke, *The Times*

'Engaging, closely-observed and beautiful . . . this author's deep love of the world around him is as *inspiring* as it is *entertaining*.'
Bel Mooney, *Daily Mail*

'My holiday reading: John Lewis-Stempel knows not only all about the different kinds of life in such a place and how they all fit together, but can also *write so vividly*.'

Philip Pullman, *Guardian*

'*My book of the year.* Each *beautifully crafted* sentence provides a stepping-stone to absorb and understand the land, with the writer's lyrical voice acting as guide and trusty staff as well as *illuminating* the mind's eye with wonderful imagery and perceptive literary devices.'

Stuart Winter, *Sunday Express*

'John Lewis-Stempel bravely relies on lively elegant prose. His *thoughtful*, *discursive*, often *humorous* and always enjoyable narrative conveys a vital message. . . A worthy successor to the man who is clearly his mentor, the late nineteenth-century countryside writer Richard Jefferies.'

John Akeroyd, *Spectator*

'A *beautifully written* and *insightful* diary with fascinating historical and literary diversions.'

Brett Westwood, *Countryfile*

'Once you become immersed in [John Lewis-Stempel's] *joyful* encounters with a host of wild flowers and creatures in his patch of Herefordshire, it's hard not to want to spend time – a lifetime at least – in his meadow.'

Gardens Illustrated

'His keen observations of the daily life of a meadow echo the spirit of BB . . . a *divinely earthy study.*'

The Field

'An *insightful* meditation on our relationship with nature, and on nature writing itself. *5 stars.*'

The Lady

'It's the *vignettes* that suck us in . . . It's enough to make any reader want their own meadowland.'

Western Mail

'Arresting, *warm* and delightfully well observed. The result is a worthy *tribute* to a fragile ecosystem.'

Jules Howard, *BBC Wildlife Magazine*

'Finely observed, and very finely told, observations of nature and also of the work and worries associated with managing this small plot of land. It's *intelligent*, *delightful*, *cultured* and *informative* . . . I recommend this book very strongly.'

Mark Avery, environmentalist and author of *Fighting for Birds*

'[John Lewis-Stempel's] descriptions are *intimate* and *lyrical*, giving an overwhelming impression of an author who cares deeply for the landscape in which he finds himself . . . We are in a golden age of nature writing, and John Lewis-Stempel [is] among its finest exponents.'

The Tablet

'Seen through the eyes and pen of a writer *engaged* in wresting a living from the land, yet also alert to the silent signs of the seasons . . . *paints a vivid picture.*'

Resurgence & Ecologist

'A *wonderful, intimate* recording of the passing months and seasons of an ancient meadow . . . It encourages us to cherish, to look and to experience nature more closely.'

Buzz About Bees

'A *poet's pen.*'

Garden Design

'[A] *beautiful* book . . . A vivid day-by-day account of the wild things he witnesses.'

Tim Dee, *Country Life*

'*Absolutely magical.* A marvellous read.'

Jilly Cooper

'A *wonderfully evocative* and restful read.'

Dame Athene Donald, *Times Higher Education*'s Books of 2014

'One book shone out for its *unaffected simplicity* and *grace*: *Meadowland: The Private Life of an English Field.*'

Hugh Thomson, author of *The Green Road into the Trees*

MEADOWLAND

The Private Life of an English Field

John Lewis-Stempel

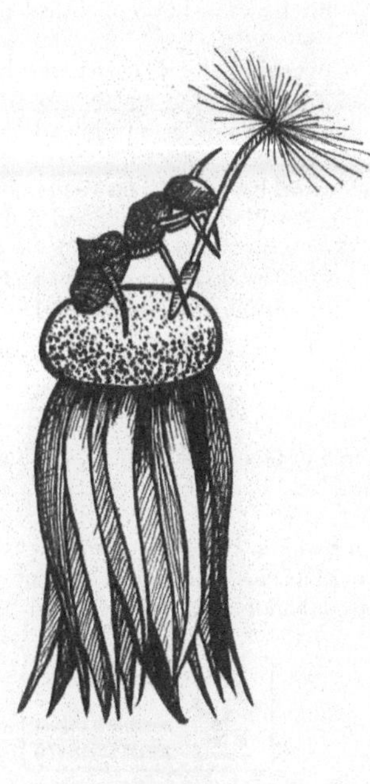

BLACK SWAN

TRANSWORLD PUBLISHERS
61–63 Uxbridge Road, London W5 5SA
www.transworldbooks.co.uk

Transworld is part of the Penguin Random House group of companies
whose addresses can be found at global.penguinrandomhouse.com

First published in Great Britain
in 2014 by Doubleday
an imprint of Transworld Publishers
Black Swan edition published 2015

Illustrations and map by Micaela Alcaino

A CIP catalogue record for this book
is available from the British Library.

ISBN
9780552778992

Typeset in 12.5/16.5 pt Goudy Old Style by Falcon Oast Graphic Art Ltd.
Printed and bound by Clays Ltd, Elcograf S.p.A

Penguin Random House is committed to a sustainable
future for our business, our readers and our planet. This book
is made from Forest Stewardship Council® certified paper.

For Penny, Tristram and Freda. Of course.

CONTENTS

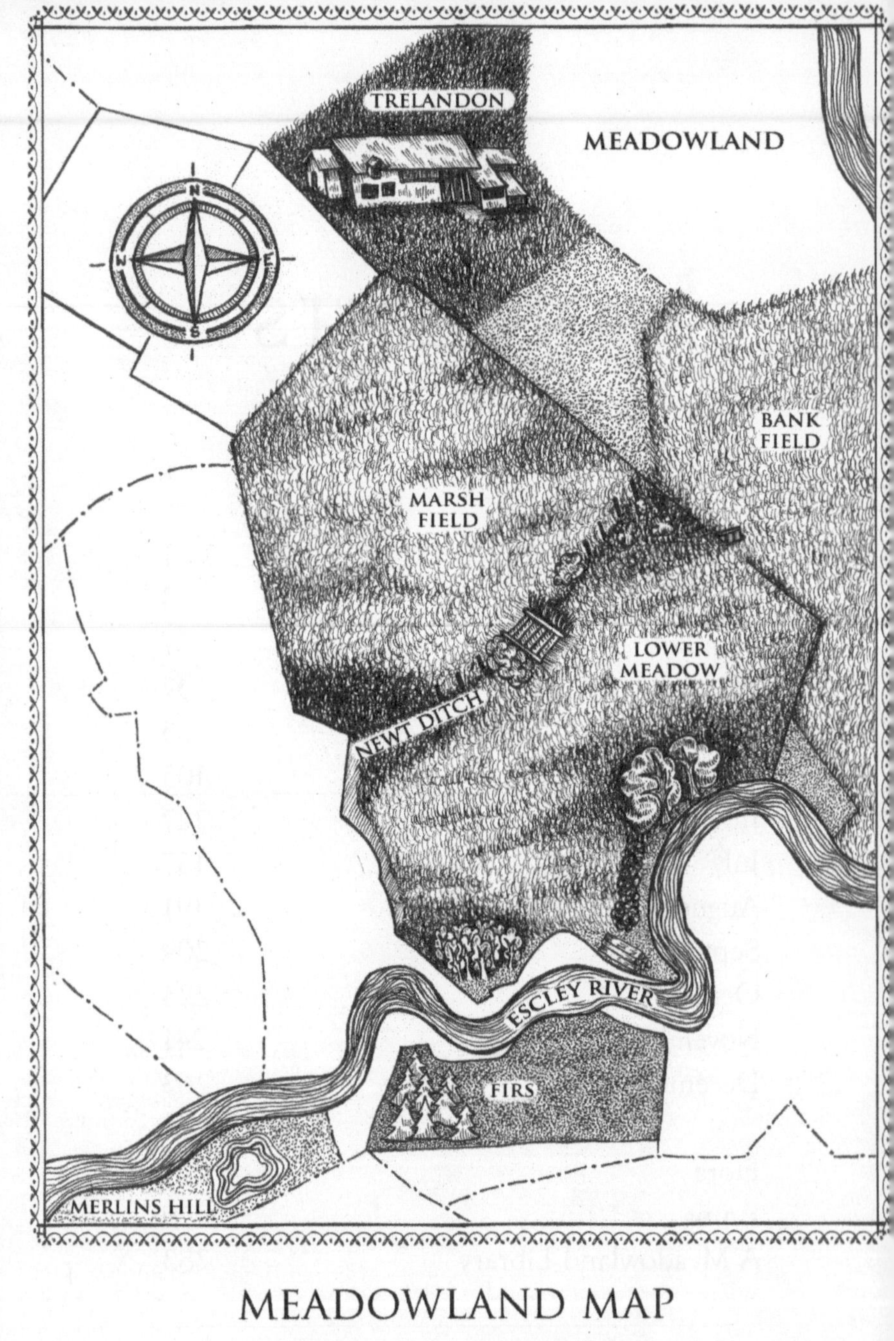

MEADOWLAND MAP

PREFACE

I can only tell you how it felt. How it was to work and watch a field and be connected to everything that was in it, and ever had been. To rationalize it . . . is pointless. The Romantic poet William Wordsworth was not always the most reliable recorder of the British countryside, but this he got right:

Sweet is the lore which Nature brings;
Our meddling intellect
Mis-shapes the beauteous forms of things
We murder to dissect.

JANUARY

Meadow pipit

THE ICE MOON is already rising over Merlin's Hill as I go down to the field at late evening to watch for snipe. There is real cold on the back edge of the wind, which rattles the dead tin-foil leaves left clinging on the river oaks. As I open the gate, my heart performs its usual little leap at the magnificence of the view: the great flatness of the field, its picture-frame of hedgerows, the sloping smoothness of Merlin's Hill to the left, then right around me the forbidding dam wall of the Black Mountains. There is snow along the top of the mountains, snow as smooth as wedding cake.

Stepping into the field is to step on to a vast square stage in which I am the last person on earth. There is not a house or person or car to be seen. It is the sort of field where, as you step in, you breathe out.

The snipe like the wet corner of the meadow, where the old ditch is broken, leaking out its contents, and where sharp sprigs of sedge have taken hegemony. The snipe have come in here late for the two nights past, where the ground is amiable to their dagger beaks and the sedge offers shelter.

Frost already spectres the grass of the field. A small flock of brown meadow pipits rise up in front of me, as though hesitantly climbing invisible stairs, chattering as they go. The nondescript meadow pipit is gregarious in winter, and is a true bird of grassland. The bird's Latin name is *Anthus pratensis*; *pratensis* is Latin for 'of a meadow'. 'Pipit' is for the bird's piping song; then again the bird is also known as cheeper, teetan and peeper in verbal reproduction of its call. All of which show how impossible it is to represent the complexity of birdsong in mere human words.

I slither down into the ditch at the far side of the field, which borders Grove Farm. This is the far west of Herefordshire, where England runs out, and the rain falls. This ditch, built to take the run-off from the fields above, is deep enough to have served a soldier in Flanders a century ago.

In the seeping red-walled ditch I wait with my arms propped on the top. I like waiting in the ditch, invisible. Sometimes I bring my shotgun, to shoot pigeons, pheasants and rabbits, but not snipe. The diminutive wader with the stiletto on its face is too rare a visitor to kill by my hand. It would be like murdering guests. A blackbird spinks in a far-away hedge.

The snipe do not come. But snipe are always mysterious; their plumage is sorcery, a camouflage of earth-blending bars and flecks. After about forty minutes when I am old and stiff with cold, and about to clamber up the ditch side, I see, from the corner of my eye, a dim shape pushing under the fence wire, leaving yet more of his silver bristles on the bottom strand of barbs.

We attribute almost supernatural olfactory powers to animals, but the truth is with the wind blowing towards me he is oblivious to my presence.

I recognize him as he advances into the field by his dragging back leg. It's the old boar badger. Badgers do not truly hibernate but he has been underground for days, avoiding the searing hoar frosts. Brock is a Nazi, a follower of Goering's maxim 'Guns Before Butter'. Although he must be hungry, he chooses first to patrol his territory.

Amusingly, the eastern boundary of his territory is the same as ours; he has adopted the human's stock fence as his national border. Along this the badger now shambles, black-and-white snout to the ground, stopping every five yards to squat and scent. The sun has long since perished, and in the quarter moonlight I can only make out his progress by the startling luminescence of the white bands on his head.

Satisfied with his noisome defences, he starts to haul himself across the field towards me.

For a sizeable mammal, badgers like the smallest morsel to eat. When he is within twenty yards I see he is flipping over old cow pats, with all the aplomb of a pizza chef. In this cold there can be few worms, but in late summer, when the grass has been cut for hay and it has rained lightly, I have seen the whole badger family out hoovering up earthworms by the hundred. A badger can easily eat 20,000 earthworms in a year. But then this 5.7-acre field probably hosts 6,000,000 *Lumbricus terrestris*; the badgers are unlikely to run out.

Tonight, however, the pickings are poor and he shuffles off. I follow his lead. Which is as it should be. He has primacy. The badger is the oldest landowner in Britain, and roamed the deciduous forests of southern England long before the Channel cut us from the 'Continent'. On the way across the field I push over some cow pats with my wellingtoned foot to see what the badger was eating. Small, glistening grey slugs.

I was not right to say the field was flat, although it is unusually flat for hill country. The field has a gentle tilt, west to east. At first glance, like all fields, it seems one habitat, but like nearly all fields it is more than one. Look again. At the two gateways where the cows stand and stare, the ground is bare, making scars in the gathering moonlight. Where the western ditch, which takes all the water from the Marsh Field above, leaks, the ground is going to bog and snipe. Part of this ditch sweeps into the field and is deep and slow enough to be a rectangular pond; it is here that the frogs and newts breed. A finger of the field sticks untidily out and is walled by trees, and is never cut because there is no room to get a tractor (or, years ago, a horse) and mower in. Under most of the hedges which ring the field, the ground is dry, especially the north end of the west hedge; here the sheep like to sleep and shelter, leaving twists of their fleece in the hawthorns and their black-green dung pellets on the ground. They are doing so now, a flock of thirty Ryelands, fifteen Shetlands and ten Hebrideans, breathily chewing the cud. It is here the thistles grow, and the brilliant goldfinches descend in their charms to feed on the October seed heads.

Lie down for a floor-across view in the frost, and the grey field is not so smooth after all, but bears the bumps and pockmarks of centuries of use. An arterial network of paths spreads across and just discernibly dents it, the trails of generations of sheep. Hoof marks from last year's cattle have collected water, reflecting the moonlight, as though someone has scattered hundreds of pocket mirrors.

The field has unseen contours too. Getting back up, I find the invisible point in the middle of the field where the air temperature changes, enough to make me shiver.

A narrow mountain river runs alongside the eastern edge of the field, finding its way to the sea. Over gravel shillets and into glass pools, and round an arching loop to leave the promontory, or finger as we call it. Most of the bankside is steep and covered in a thicket of holly, alder, hawthorn, hazel, field maple, ivy; it is the overgrown child of an antique parent hedge. There are two kingly oaks in the thicket, which in their dotage clutch the river bank, to lean precipitously over the river, on roots elephant-trunk thick and which swirl thrashingly into the ground, leaving nightmarish, dark troll holes. The oaks, which are around seven hundred years old, are remnants from the time when this isolated valley was wooded.

Where the river leaves the field, the thicket grows out into a small copse; here, secreted in the bracken and scrub, is a fox's earth. Foxes like to make their homes near water.

The river has a name, the Escley. In his *The Place Names of Herefordshire* of 1916 the Reverend A. T. Bannister advised about the noun 'Escley': 'Wise students refuse to discuss river-names; but one is tempted to connect the word with the Celtic root from which comes Exe, Usk, Ock, and Ax-ona.' He is likely correct on the derivation, which would seem to come from a Brythonic root word meaning 'abounding in fish'. Put more plainly, Escley is related to the Welsh for fish, a lexical reminder that national ownership of this borderland ebbed in the past, and the field today is just a mile inside England. The Escley does indeed have fish, and the patient person who can cast flies between the alders that line the river's route may find trout. Tonight the Escley is chattering discreetly.

As I leave the field the raven croaks, despite it being nighttime, to remind me that a pair of this species nest and roost in a small grove of fir trees just across the river and they have the best view of the field of all. Ravens mate for life, and this pair has been here since we have.

When we moved to this farm, the field was my delight and my despair. No field has a finer aspect; only if I spin the full 360 degrees can I see houses, and then only three of them, one of which is ours. Such was the joy; the horror was the state of the sward. My head was stuck in conventional farming thinking, and I deplored the lack of clover for our cattle and sheep, and two patches of the field were devastated by wireworm.

It became the field I did nothing with, the place I plonked livestock when nothing better was available. But then, nobody else had done much with the field either; there were stands of thistles of a density which suggested remarkable ancestry.

Sometimes neglect is good. In the city the rich folks live on the hill. In the country it's the poor folk. The big beef farmers and the corn barons have the flat land. Hill farmers are frequently too capital-lacking to make big changes to the landscape. Or spray gallons of herbicide on to it. Nothing conserves like poverty. One summer I let the field go, instead of shuffling livestock on to it.

The peasant poet John Clare called plants 'green memorials'. By late June the field had sprouted flowers I'd forgotten existed, flowers such as knapweed and bugle, which were testament to an agricultural usage other than animal parking lot.

Once upon a time the field had been a hay meadow.

W

7 JANUARY Snow settles its medieval quiet on the land. Snow five inches deep, deep enough to sledge on, and Tris and Freda make a Cresta run down neighbouring Bank Field. Away in the unseen village other children's voices shriek delight.

I too use a sledge, though not for such fun as hammering downhill at too many miles an hour. I tie a bale of bright hay to a storybook wooden sledge and haul it down to the field in a manner I romantically and vaingloriously imagine to be that of Scott of the Antarctic. The sheep are more interested in the accompanying half-sack of sugar beet, and crowd round me, the fearless and the hungry jumping up. The snow hangs in a deb's delight of pearl beads around their necks. The snow suppresses the herbid smells of nature, and the musk of sheep is unbridled.

Vast shining tracts of the field are unspoiled by the sheep's feet. In these virgin spaces I feel as though I am exploring a new planet, which of course I am. The white planet. So cruelly bright is the sun off the quartz snow that I have to squint as though peering into the future.

Actually, not all of the field is white; there are two patches of green at the centre, where water oozing out of the ground has prevented the snow from settling. They are oases around which birds have flocked to drink and probe the ground for food. In the soft mud I can see the footprints of pheasants, and also the faint triangular marks of a smaller long-toed bird.

They come back briefly in the afternoon. Three lapwings.

Farming has changed beyond all recognition in the last seventy years. Population pressure means that farmers need to grow more, more quickly. In the 1930s, Britain's farmers produced enough food to feed 16 million people; today they produce enough to feed 40 million people. Almost all grassland is now managed 'intensively', whereby selected species of grass have fertilizer and herbicides applied to them. On intensively cultivated grassland, grass yield has gone up by 150 per cent since the 1940s.

There is a cost. Ninety-seven per cent of traditional meadows have disappeared. Treatment with artificial fertilizers did not benefit the more fragile meadow grasses, and the more vigorous types smothered them to death. Neither did the new regime of cutting for cured grass (silage) two or even three times a year help, because the first cut came in May before flower seeds had set, and while ground-nesting birds and mammals were still rearing their young. Some animal species have become extinct. I have not seen a corncrake since the 1970s, when I almost stepped on one in a hayfield when childishly shooting rats with an air rifle. The bird is extinct in England. It stands as a symbol of the untold damage we have wreaked on the British landscape and our natural heritage, through the drive to produce as much food as possible on our crowded little island.

Many of the plants in a traditional meadow, which was cut late and then grazed by small numbers of cattle and sheep, did not have a direct agricultural food value, but they preserved the balance of nutrients, and provided wildlife with sustenance. Didn't they also make Britain special? When khaki men on the Somme or in Burmese jungles thought of their homeland, did they not picture wildflower-strewn meadows, with cottages and rolling hills?

'Meadow' is surprisingly strict in its meaning, and is from the Old English *mœdwe*, being related to *māwan* – to mow. A meadow is a place where grass and flowers are grown for hay, the dry winter fodder for livestock. A meadow is not a natural habitat; it is a relationship between nature, man and beast. At its best, it is also equilibrium, artistry.

9 JANUARY More snow comes, and a west wind with it, blowing the snow into ridges; the effect is as if a white tidal sea has been over the field and withdrawn. Some scruffy, wizened spikes of thistle pierce the snow, spoiling the illusion.

The sheep paw at the sparkling spectacle for the grass underneath, and lurk for hours around the hayrack. The Shetlands gnaw at thick ivy tendrils in the hedge, back to the white bone, leaving the skin to weep orange at its edges. They have also pulled at the leaves of the bramble, a deciduous plant with a surprising tendency to the evergreen. The temperature reaches minus seven at night, and a sheet of ice forms over the slow stretches of the Escley. I watch it grow, a creeping, pale fungal invasion. The water in the pond is inch-thick plate glass. It groans and protests though it will bear my weight.

By the Grove ditch I can see the wide paw marks of the badger, and where his shaggy coat has scuffed the snow's surface. In the bank behind the Grove ditch there is a small rabbit warren; usually the rabbits graze into the Grove field, and use the burrows on our side as escape hatches. This morning they have come through into the field and scraped to get the grass under the sheltered side of Marsh Field hedge. Rabbit's scientific name is *Oryctolagus cuniculus*, the 'digging hare'. But the snow is deep and the ground iron; the rabbits have also been up on their back legs nibbling the sweet hazel bark.

The children are at school. The sheep are too defeated by the cold to bother baaing. Only the leathery creak of my feet in the snow disturbs the world. Even the river is quiet.

There was once a sea over this land, and fishes swam in the meadow. The meadow then was south of the equator. During the early Devonian period, 425 million years ago, the place I am now standing in was covered by a shallow tropical estuary, the bed of which writhed with primitive fish and crustaceans – acanthodians, eurypterids, cephalaspids, pteraspids and the scientific like. I know exactly what would have swum around my feet because an old farm quarry half a mile upstream, Wayne Herbert, has divulged hundreds of fossil remains from its green siltstone, including one of an early lamprey, which was wonderfully named *Errivaspis waynensis* in honour of its place of finding. The same green siltstone lenticle in which *Errivaspis waynensis* was discovered runs under the field, whose geology is easy to determine: the river has cut away the side of the field, so a cross-section is handily on show. There is of course a good reason why hay meadows are traditionally located next to rivers; if grass is to grow lush and thick it needs all the water it can get. The river obliges by seeping its bounty into the depths of the field's soil through trillions of subterranean capillaries.

Standing in the river looking at the bank, I can see 450 million years of geological history before me, with the horizontal green layers of siltstone lenticle at the bottom. I spend the day chipping away with a chisel and hammer and, even in this bleak midwinter, it becomes warm work.

Some time in the mid-afternoon, when a shard of sunlight plunges through the oak branches on to the green slab, I find what I have come for. A fragment of fossilized fish scales, almost certainly from an acanthodian. These 12-inch, heavily scaled vertebrates were the first jawed fish, the direct ancestor of the trout, loach and bullheads that live in the river today. And of the curious minnows that are investigating my wellingtons in the pellucid water.

On a drive into Hereford I take the cross-country route, hoping to avoid the traffic bottleneck. (Some hope.) Despite the sparsity of population, I count at least five houses where the inhabitants have strimmed their roadside verge to within a centimetre of its life. Internally I rail at the suburbanity of such an aesthetic (why move to the country if you want to turn it into Hyacinth Bucket's Blossom Avenue?), and rather more honourably deplore the ecological holocaust. Roadside verges are often remnants of ancient meadow – and in some areas, the only remnants of ancient meadow – and are flora rich, and the sanctuary of wild animals. There is a rust-covered kestrel hovering over an uncut swathe at Wormelow.

11 JANUARY The snow still straggles and streaks the ground. Such is the mercurial nature of January that the wind drops, the sun startles between clouds, and midges dance in the columns of light. My most faithful companions in the field, the meadow pipits, lark about. I do not dance; there is something oddly unhealthy about balm in January.

By mid-afternoon the snow is melting rapidly. The soil under the grass is already saturated.

According to the Agricultural Research Council's *Soil Survey of Great Britain*, Bulletin No. 2, 1964, the geology of the field is 'Devonian marl with fine grained sandstone bands and, very locally, thin drift'.

I prod my fingers through the slushy grass into the soil and clutch a handful of this Devonian marl. To my hands, and those who work it, the soil of Herefordshire is thick, red, cold clay. Squeezed and balled in my hand, it rolls solid into a mini Earth.

Nature abhors easy classification. Strictly, the field is 'neutral' unimproved grassland, meaning the clay soil is neither strongly acidic nor alkaline, with a pH of around 7; actually parts of the field err to acid – pH 4.9 to 5.4. To the common-or-field botanist like me, this means the field is the abode of acid-thirsting plants such as sorrel, and I love sorrel, with its misty-red tops in summer, and its lance-leaves redolent with minerals and vitamins for beast and man.

Water does not easily drain through this Devonian marl which is neutral going on acid.

By the next day much of the field is glistening deep in an inch of water, and I have to move the sheep out temporarily. Water is seeping in a continuous sheet off the field into the river. The earth sucks at my feet, making my gait arthritically unsteady.

14 JANUARY Overnight the Escley breaches its bank to flood across the bottom of the finger. By the time I go down at midday, the water has subsided, leaving a catastrophic carpet of broken branches, logs, trunks

and twigs. In the copse I can see that the flood has just missed the fox earth.

The level of the Escley may have gone down, but it is still roiling with wild sea-fury.

Late evening in the field; the unseasonal warmth and wet has pushed earthworms up to the surface, but they drown anyway in the pewter pools, each worm a silent white S. In a cloisters gloom I can see a fox paddling along, lapping up worms galore. The fox and that pioneering naturalist Gilbert White would be of one accord on earthworms: 'though in appearance a small and despicable link in the chain of nature, yet, if lost, would make a lamentable chasm. Worms seem to be the great promoters of vegetation, which would proceed but lamely without them.'

Earthworms are not, generally, very active in winter, and leave the work of cultivation to the worms *Allolobophora nocturna* and *Allolobophora longa*. (And you thought all worms were the same.) Meanwhile, I take solace from the old weather saw:

The grass that grows in Janiveer
Grows no more all the year.

15 JANUARY Snoopy, our miniature tri-coloured Jack Russell, yaps at something under the goat willow,

whose wands are aerials into the descending mist. When I get to the dog his nose is pierced with blood, which is also speckling his Chippendale front legs. He has attacked a hedgehog, which, confused by this faux spring, has lumbered out of hibernation. Rolled tight in a ball on the foul leaves, only a slight breathing reflex betrays that nature's giant pin-cushion is alive.

Some stupid impulse makes me touch the hedgehog's prickles to evaluate their sharpness. The trademark spines, all five thousand of them, can grow up to 2.5cm long. Kneeling, I go slightly off balance and apply more pressure than intended. A spine goes under my fingernail. Blooded, both dog and I withdraw; the defensive thorns keep most predators, as well as me, at bay. A badger, though, will unroll the hedgehog and eat it from the inside, discarding the coat as wrapping.

Behind me the river shouts with the abandon of a football crowd.

17 JANUARY Old Twelfth Night. I succumb to a bad case of tradition and go wassailing. Wassail is derived from the Middle English *waes hael*, meaning good health. In 'social wassailing', one has a drink with one's neighbours; in the cider-producing counties of the west of England, there is also 'orchard-wassailing',

where the apple trees are awoken by being beaten with sticks, a piece of toast placed in the branches, and cider sprinkled around the roots. All to ensure a good crop in the year ahead. Ideally, one should sing a wassail song. Something like:

Wassail the trees, that they may bear
You many a plum, and many a pear.

Wassail might be Middle English by way of Old Norse, but the custom is probably older, pre-Christian, even a relic of the sacrifice made to Pomona, the Roman goddess of fruits. Wassailing used to be big here in Herefordshire. According to *The Gentleman's Magazine*, February 1791:

> In Herefordshire, at the approach of the evening, the farmers with their friends and servants meet together, and about six o'clock walk out to a field . . . In the highest part of the ground, twelve small fires, and one large one, are lighted up. The attendants, headed by the master of the family, pledge the company in old cider, which circulates freely on these occasions. A circle is formed around the large fire, when a general shout or hallooing takes place, which you hear answered from all the adjacent villages and fields. Sometimes fifty or sixty of these fires may be seen all at once.

The fires represent the Saviour and His apostles.

Mysteriously, Penny and the children are too busy to join me in the wassailing, so it is just me, a piece of bread, a bottle of Westons, the shotgun and my black Labrador, Edith.

In groping blackness I toast the two vintage apple trees in the river bottom of Bank Field with the toast and the cider. Then like a vandal I fire off the shotgun into the dim treetops, to scare away the spirits.

If both barrels from a 12-bore do not deter malignant auras, nothing will.

The old ways do not seem so mad in an ancient landscape where I can barely see one electric light, and I can hold in my cupped hand the eternal peace of night. In these foothills of the Black Mountains more than half the farms have most of their historic footprint, and the small hedged fields result from medieval woodland clearance. Such as this field, such as the hay meadow.

W

18 JANUARY The weather is sly. From behind windows it is overcast, damp, nothing special. I have let the sheep back into the field; only when I am halfway down to the field carrying a plastic tub of mineral lick to them do I feel the knife go through my coat and try to hollow out my being. My hands (I have

stupidly mislaid both my gloves and my spare gloves) are blackberry blobs; I pass a great tit, its eyes misted with hopelessness, lying on the wan grass under the Marsh Field hedge which is naked and useless and no home at all. The winter is taking its tithe; I determine to pick up the bird on my way home, and warm it and feed it.

There is no rain; even so the wind sculpts raindrops from my own eyes, so everything is opaque, fish-lensed, underwater. By the time I have deposited the mineral bucket and got back to the great tit, it has died. I hold it in my numb fist; it weighs nothing. The white patches under its eyes look like cartoon tears.

There is a kind of life in death. The winds, snows and floods of winter have scraped the countryside clean, ready for a new start. The last lingering leaves on the oak are down, trees and hedges are X-rays of their former selves, and the two grey squirrel dreys, one in the hazel in the copse and one in the pollarded oak in Marsh Field hedge, are blots in the fretwork of branches. A crowd of starlings moves along the fields in an avian Mexican wave. I like their company on this barren, skeletal day, when the only other sound is the pitiless mewing of a circling buzzard.

The snow has not finished with us. All night, flakes flitter down, so the snow is fully three inches deep by afternoon, when the wind crisps its surface and the velvety rabbit noses of the ash buds seem the only soft things in the field. Along the rim of Grove ditch there are the tinkling paw marks of a weasel or stoat; then the signs of a flurrying gallop, a struggle, spots of brightening blood, then the broad scuff mark where the rabbit body was hauled into the hedge.

Small is the difference between the pad marks of the weasel and the stoat; I settle on stoat because of the length of the stride, the stoat being the bigger of the two *Mustelidae* cousins.

For an hour in the afternoon I sit on an empty plastic sugar-beet sack in the corner of the field where the hazel hedge has broken down. The liquorice aroma from the beet remains is childishly comforting. I am so intent on looking at the scene of the crime across the field that I fail to spot the stoat sitting up staring at me from five yards to my left. A stoat it is; weasels do not turn white in winter. A rather patchy white to be sure, with a brown blotch on the flank and shoulder.

I blink first, and the stoat lopes away, sending up small blizzards of snow.

Over in the copse a robin sings fortissimo. There is a wren, no bigger than a moth, working the leaf litter in the hedge along from me; wrens do not sing in midwinter, they are too busy foraging.

The fog comes down, and erases all the world beyond the field. The field is an island.

We are snowbound. To get up the farm track to the lane requires clearing a path with a shovel on the front loader of the tractor.

I resume the sledging of hay and beet pellets down to the sheep in the field; around their troughs and hayracks they have trodden the snow down into the mud. A grateful blackbird is probing the exposed earth, and house sparrows are seeking the vestiges of grass seed heads and beet bits in the bottom of the covered feeder.

Another blackbird is pecking at the globe of mistletoe hanging in the gateway hazels; most of the farm's thrushes, together with the migrant redwings and fieldfares, have retreated before the west winds and gone to the villages and lowlands. The sea-wrack leaves of the mistletoe give the plant the appearance of being stranded by an invisible oceanic flood.

At night a fox somewhere across the river yips at the moon. I am unable to resist a go down the Cresta run; the slough of the sledge is the only slander in the moonlight.

W

21 JANUARY I count Edward Thomas among my favourite poets. In fact, 'Adlestrop' is one of the only two poems I know by heart. (The other being Shelley's 'Masque of Anarchy', learned in a punky rebellious

teenage phase.) When Thomas was asked by Robert Frost why, at the age of thirty-five, he was going off to fight in the First World War, he bent down and kissed the earth of England. 'Literally, for this,' he said. I would do the same if asked. Thomas thought the greatest gift he could give his children would be the English countryside. In 'Household Poems' he wrote that his bequest for his son Merfyn was:

If I were to own this countryside
As far as a man could ride,
And the Tyes were mine for giving or letting, –
Wingle Tye and Margaretting
Tye, – and Skreens, Gooshays, and Cockerells,
Shellow, Rochetts, Bandish, and Pickerells,
Martins, Lambkins, and Lillyputs,
Their copses, ponds, roads, and ruts,
Fields where plough-horses steam and plovers
Fling and whimper, hedges that lovers
Love, and orchards, shrubberies, walls
Where the sun untroubled by north wind falls,
And single trees where the thrush sings well
His proverbs untranslatable,
I would give them all to my son
If he would let me any one
For a song, a blackbird's song, at dawn.
. . .
Then unless I could pay, for rent, a song

As sweet as a blackbird's, and as long –
No more – he should have the house, not I:
Margaretting or Wingle Tye,
Or it might be Skreens, Gooshays, or Cockerells,
Shellow, Rochetts, Bandish, or Pickerells,
Martins, Lambkins, or Lillyputs,
Should be his till the cart tracks had no ruts.

Field names rarely match the romance of village names (Wingle Tye!, Margaretting!); hardly ever do they lift themselves above the ultra-prosaic. We always called the meadow 'Copse Field' or 'Finger Field'; the neighbours on arrival told us it was 'Bottom Field'. I take advantage of a dreary day to look up records in Hereford Reference Library, where some assiduous and civic-minded amateur historians have compiled a volume of local field names. Almost alone among the books – everyone under forty being on a computer – I uncover the field's historic, official title, as given by the Tithe Survey of 1840. Lower Meadow. This is next to Bank Field. Nearby fields include Big Field, Sheep Shed Field, Long Pasture, Cow Pasture, Eight Acres, Field Down the Road, Far Field, Big Meadow and Flat Field.

Field names are not the only uninspired descriptors in the English countryside. Farm names are invariably utilitarian, as the same source confirms. The nadir resides about four miles away. Farmhouse Farm.

The tithe survey was carried out following the passing of the Tithe Commutation Act of 1836, which was designed to rationalize the system of financial support for parish priests, which by that date had become bogged in confusion and evasion. The act was designed to codify formally the practice of paying tithes in cash (the 'commutation' of tithes), rather than in animals or agricultural produce, and was based on the amount of land which people owned. For the system to be effective, maps had to be produced, together with lists of landowners and how much land they held. Looking at the maps though, I realize what a memorial field names are. Any field name that includes 'Stubbs' or 'Stocking' refers to it having been cleared from woodland, 'Butts' may well have been the place the medieval locals practised archery, and 'Walk' indicates land formerly given to the common grazing ('walking') of sheep.

The field names also crystallize Herefordshire dialect: 'The Tumpy' is a bumpy or steep field, a 'tump' being the vernacular for hill. And surely 'Sour Meadow' was a place of bad grazing? You can see too the pattern of former village settlement: 'Butcher's Shop' was adjoining the local meat emporium, long since gone.

One name sits on the Tithe Map with the mien of a gravestone: 'Cuckoo Patch'. There are hardly any cuckoos in the valley now.

People needed to know field names, which were their places of work. Children and wives needed to know where to take men their 'elevenses' and 'fourses', their cider or tea, their bread and cheese.

Outside, a seeping rain is still coming down, so I vote for warmth and rummage along the shelves of the local history section, where I find a reproduction of the 1664 Militia Returns for the village. (The returns were a form of taxation.) There listed halfway down the page is one Sam Landon, liable to pay tax on £6 of income.

It is Sam Landon from whom the farm takes its name, Trelandon, being Welsh for house of the Landon family. He rented from the co-heirs of 'Ye Lorde Hopton'; the Hoptons would own the farm for another hundred years, until they sold to the Marquis of Abergavenny. The Marquis's family retained the farm until 1921, when they, like so many other landowners in the shadow of the Great War, sold up. Between 1918 and 1922 a quarter of the land mass of Britain changed hands, including Lower Meadow. It was a sale of land unprecedented since the dissolution of the monasteries.

When Sam Landon took on the farm he brought with him a newfangled idea, which was to live on site. Previously, the prevailing pattern of rural settlement was that all toilers of the land, labourer and farmer alike, would live in a village and walk to work. There was nothing quaint about Herefordshire medieval villages, which largely consisted of lots of tumbledown hovels of sticks and clay, in which yokels insisted on burning elder wood, and wondered why they died at night. (Burning elder releases cyanide.) So poor were country people in these valleys under the Black Mountains that one seventeenth-century local gentleman, Rowland Vaughan, declared them 'the plentifullest place of poore in the Kingdome . . . I have seene three hundred Leazers or Gleaners in one Gentleman's cornfield at once'; the great impoverished were scratching around in the dirt for the 'gleanings', the left-over grains.

One assumes Sam Landon was glad to leave the madding throng and build his own house in splendid isolation. He was certainly a man of notions. By Hooper's rule of dating hedges (age = number of species in a 30-yard stretch × 110 + 30) I have estimated the western hedge of Lower Meadow to be 350 years old. This hedge divided off Lower Meadow from the wetland above it; the ditch dug at the same time dried the bottom land. More, by dividing off Lower Meadow, Landon was able to stop stock grazing this drier, better land over the spring and summer.

He turned a field into a hay meadow proper.

Of course, rummaging around in forgotten documents entails the same risk as going through someone's diary. You may discover information you had no wish to know. A flick through the pages of a book on Herefordshire informs me that the rainfall on this far western edge next to Wales averages 30–40 inches a year. Amusingly, the same shelf has the history of the school my paternal grandfather attended, and which is also situated bang on the English–Welsh border. In the eighteenth century this establishment advertised in London for boys to come and learn Latin and Greek in the temperate, healthy climate of Herefordshire's borderland.

I almost laugh aloud.

The rain is still pouring uncontrollably from the sky, and the library is closing for the night. When I return home, I tog up in my Barbour, battening down every zip and button, and slop down to Lower Meadow, my shoulder a prow into the spray. The clay base of the field is again saturated, and holding a good half-inch of water on its surface. It is dark, the way only a January night in a rainy, lightless valley in the middle of nowhere can be dark. Thick, cellar-dark. I cannot see the water on the surface of the earth. I can tell its depth by the plash of my wellingtons.

W

24 JANUARY No bird is less of the meadow than the kingfisher; in five years I have never seen it deviate in its flight from the river bed, its sole route-finder. But it

is often there in the periphery of the field and of my vision, a neon-turquoise spark, which leaves an iridescent flash in the atmosphere, to die away slowly. A radioactive particle decaying.

Now the kingfisher comes, flying on a perfect horizontal plane, suspended equally between the river and the sky.

This is the halcyon bird of mythology, which allegedly laid its eggs on the ocean, an act by which the sea was calmed. Hence the 'halcyon days' of poets and Shakespeare. Some believed that a kingfisher could divine the weather as well as determine it; a dead kingfisher, suspended by its head, would turn its beak to the wind like a multicoloured weathercock.

A field is a landscape is a soundscape. The 'zeep zeep' of the kingfisher is an occasional contribution at the edge of consciousness.

I'm in the promontory, where the alder logs are spread with a caramel layer of velvet shank mushrooms. *Flammulina velutipes* is one of the few mushrooms that fruits in winter; it is also edible. In Japan it is the prized culinary delight *Enokitake*. Only the Jew's ears on the elders which are lolling an arm's length into the field rival the velvet shanks for hardiness.

W

26 JANUARY I'm out when Roy Phillips the contractor comes to cut the farm's hedges with a flail cutter on the back of his Ford County tractor. The hedges should be 'laid', slit and woven by hand, but this is one of those time-consuming jobs that is on the end of an endless list. Because Lower Meadow is suppurating water Roy has only been able to get the tractor into part of the field; only one and a half hedges have been trimmed, giving the field a dissolute, half-shaven look. The flailed hedges are stark and square. Scalped to the skull. The serpentine ivy seems to be the only thing holding the bushes together. On the uncut hedges, drooping claws of alder catkins and bunches of plum ivy berries pose in flaunting juxtaposition. But the intoxicating melon aroma of the shorn hedges makes them beautiful to the nose.

There is evening sky to delight a shepherd. The vapour trail of a microscopic jet catches the crimson light so that the aircraft is illusorily powered by flame.

27 JANUARY Snowdrops and dog's mercury are out in the hedge along the farm track. The days are perceptibly lighter and longer.

I climb into the copse; in the centre, amid the sombre trees, the fox's earth has been undergoing renovation, and there is evidence of digging around the muddied main entrance. Aside from bits of dead animal (rabbit, song thrush) lying around, there is that unforgettable proof of fox habitation: a sour, pungent odour when one sniffs over the dungeon hole.

By now the foxes will have mated. Assuming successful implantation, the vixen will begin a gestation of about fifty-two days. Red foxes have the shortest gestation of the dog family.

On the floor of the dripping copse lies a dead blue tit, a startling jolt of colour amid the darkening leaves. I doubt if the fox, as per folklore, hypnotized the blue tit to come down from its tree. The arctic blast and the relentless after-party rain murdered many a bird.

Low winter sun comes strobing through the coppiced hazels.

FEBRUARY

Jackdaw

CANDLEMAS, 2 FEBRUARY. A morning of stultifying mist, which cloys like sweat on my face. At the far end of the field the unseen raven voices a mindless metronome of croaks. She is sitting tight on her eggs in the triangle of firs across the dull stream, and has been for the last week. Ravens are famously early nesters. Perhaps she is complaining about the badgers' housekeeping; they have done a spring clean of the sett and dumped armfuls of stale, fetid moss bedding beneath her tree.

The dew, trapped in the webs of countless money spiders, has skeined the entire field in tiny silken pocket squares, gnomes' handkerchiefs dropped in the sward. Or it has where the Ryeland ewes have not relentlessly mowed their way through the grass.

The thirty Ryelands spread out before me in the white silence are the direct descendants of the sheep that made the field in the early 1400s. Unlike the primitive sheep which dominated the English landscape until the Middle Ages, Ryelands were – are – systematic grazers of grass and not half-goats who yearn to browse trees and bushes. By their unthinking, indiscriminate eating, Ryelands suppress the more prolific grasses, and allow the more delicate grasses and flora to survive and flourish. One reason that English meadows, like Lower Meadow, boast a display range of grasses and flowers is because of Ryeland sheep. Lower Meadow has, of grasses alone, timothy, meadow fescue, cock's foot, meadow foxtail, woodrush, sweet vernal, tufted hair-grass, crested dog's tail and meadow grass. Of the grass family, the Gramineae, this is a reasonable selection in the twenty-first century; there are, though, 150 species of grass growing wild in Britain.

The thirty ewes are oblivious to my environmental design. They have eaten hard because they are heavy and round with lamb. They exude self-satisfied fecundity, which the wet lying thick on their coats does nothing to dampen. One stands square, and spins its alabaster torso so the water sprays off in a brief abdominal halo.

We have looked at one another a lot, Ryeland sheep and my family, on mornings like this over the years, because Ryelands and I are the dangling ends of dynasties long familiar to each other. My mother's maternal line, the Parrys, were the feudal stewards of Ewyas Lacy – as this valley used to be called – and helped develop the breed. The Parrys had Ryelands grazing here in these meadows under the Black Mountains five hundred years ago. And I like to think it was cousin Blanche Parry who gave Elizabeth I the stockings made from fine, white Ryeland wool which so impressed the Queen she would thereafter have no other material on her virginal legs. Blanche Parry served Elizabeth for fifty-six years as lady-in-waiting. Then again, it might have been any of the other Parrys at court who were the nominatives in the fable of the white fleece; there was Dr Henry Parry, Elizabeth's chaplain, James Parry, her Huntsman, Sir Thomas, her Cofferer, John Parry, her Clerk of the Green Cloth, Frances, her maid of honour, Katherine, her lady-in-waiting, Lady Troy, yet another of her ladies-in-waiting, or perhaps Catherine, her Lady of the Bedchamber. Or maybe it was Blanche's great-nephew, the aforementioned Rowland Vaughan, another courtier, or her cousin John Dee, the Queen's astrologer. Or perhaps it was Blanche's most illustrious cousin, Lord Burghley, Elizabeth's Chief Minister, whose ancestral seat, Allt-yr-Ynys, still stands at the lowland bottom of Ewyas Lacy valley.

The Parrys, you feel, had a bit of an armlock on the Elizabethan court.

The Parrys, though, were more than royal flunkeys. They were part of a historical phenomenon, the swing to sheep farming in Tudor times, which came about, not least, because someone noticed that grassland manured by sheep retained its fertility.

I am counting the sheep. They are all present and correct, and now it is time to go.

Candlemas, 2 February, is the day by ancient rite when the hay field is 'stopped' or 'locked up', the day when all livestock is removed.

The expectant Ryelands do not need a dog to move them out. A sheepish call of 'Sheep!' and they follow at my heels. Pavlov in his grave should applaud. The sheep know that if I call out their species, food surely follows. I do not cheat; when they are through the gate into Bank Field they find a feast of beet nuts in their long troughs.

I close the galvanized gate on Lower Meadow. Now the grass can grow unmolested until mechanical cutters come one fine summer's day. Then they will shear it down in great swathes, and it will be turned into hay for winter feeding.

Grass, the keeper of us all.

And a line comes floating back from a lesson I read in church at Christmas as a boy. Isaiah XL, vi–vii: 'All flesh is grass, and all the goodliness thereof is as the flower of the field: The grass withereth, the flower fadeth: because the spirit of the LORD bloweth upon it: surely the people is grass.'

Mooching along the winter-bare Marsh Field hedge I espy the domed nest of a long-tailed tit, an extraordinary construction of moss, plated in lichen. It looks like a verdigris egg. Or maybe a tarnished-copper Cyclops' helmet. For all their dainty artistry, the long-tailed tits have anchored the nest to the willow branchlets with stevedore stoutness. Despite my most careful disentangling, the nest bursts as I pull it free, exploding pigeon feathers from the lining as though we are having a pillow fight.

To find the archaeological artefacts of twentieth-century agriculture is not difficult. Farmers have the habit of throwing anything dead and useless out of the way, which means the cellar-dank bottom of the hedge.

I start poking with a hazel stick by the gateway. Out in the field are a handful of chaffinches, desolately picking at the sward for the stray fallen seeds of grasses and flowers. The three pairs of chaffinches that nested in the field's hedges will not wander far from them all winter. They will be joined by females from Scandinavia; the males will stay behind. The Latin name of this bird, *Fringilla coelebs*, *coelebs*, deriving from the Latin for bachelor, was given by Linnaeus, who saw only male chaffinches in his native Sweden, the females from its northern breeding grounds having flown south. Chaffinches were originally birds of the woodland. But what is a hedge other than linear woodland?

Within two minutes of hunting through the moss and layers of bog-black leaves, I find the first treasure, a long metal band, dulled by time, but not so old that it has lost all its red Massey Ferguson paint. I've removed enough of these to know one instantly; it's the guard for the belt drive on a 1970s-era baler.

As I bend down in the leaf detritus, my stick strikes a dull, tomb note: an empty brown cider flagon. 'Bulmer's Strongbow. Please Return'. There's another nearby. Then a bone-white piece of clay pipe stem.

This is the shaded, west side of the hedge. And it is exactly where we too rest while having our lunch when 'on the hay'. We have been sitting in the reclining shadows of generations of farm workers. There really is nothing new under the sun.

The field is in a mood. Sombre. Dull. The alders along the river a chill purple scrabble. Nothing to see, except three pied wagtails. A small measure of joy. The only living thing in sight.

Of course it is me that is gloomy. The field reflects the weather and the human mood.

9 FEBRUARY Blue tits whistling in the afternoon sun, a squeaky seesaw noise. Hazel catkins shagged out. Midges in Brownian motion across a golden field.

In the night I shine a mega-torch across the field, and catch the pink eyes of rabbits eating close to the Grove ditch. The rabbits are not unduly disturbed;

one is 'chinning' the ground, marking it as 'me-me' territory with the scent gland under its jaws; February is the beginning of the main rabbit mating season.

But the warren in the bank is never populous; I have never seen more than seven adult rabbits emerge from it. It is a warren for the low-status, the ex-communicated, the pariahs, and is foolishly close to the fox earth in the copse. A mere fifty yards separate them.

Like straight roads, central heating and pear trees, the rabbit came over with the Romans, and has been eating his way through the British countryside ever since. The reduction of the rabbit population – which reached 50 million by the 1950s – was the aim of the scientifically introduced disease myxomatosis. Pockets of the disease still fester locally, leaving the afflicted rabbits to bumble around with bulging, bleeding eyes, as though they emerged from the mind of Edgar Allan Poe.

But it is not 'myxy' that performs eugenics on the Grove bank rabbits; it is predation and elemental weather.

13 FEBRUARY Flurries of snow in the morning.

A raven flies over, emitting its basso profundo croak. The wingspan of a raven is four feet, and the bird always chills the land with its Gothic shadow. When I walk past the copse a jay screams at me from within. (I jump, confirming Chaucer's observation in *The Man of Law's Tale*, 'Thou janglest as a Jay'.) A wren starts up with its staccato alarm calls, the 'tecks' rattling off the hard, bare branches.

I watch it fly off into the sheltering mass of holly in the hedge. Another grateful wren joins it. And another. They are clubbing together their body warmth in a bid for life.

W

14 FEBRUARY St Valentine's Day, the day that Geoffrey Chaucer was convinced that the birds became betrothed. In his *Parlement of Foules*, written c.1381, he visioned that on this day of love the birds met and Nature:

This noble emperesse, ful of grace
Bad[e] every foul to take his owne place,
As they were woned alwey fro yer to yeere,
Seynt Valentynes day, to stonden theere . . .

Of foules every kynde
That in this world han fetheres and stature
Men myghten in that place assembled fynde
Byfore the noble goddesse Nature,
And ech of hem dide his besy cure
Benygnely to chese or for to take,
By hire acord, his formel or his make [mate].

Over the field two wood pigeons, as if on cue, fly up high with clappering wings, then glide down in school-time paper darts. This is their courtship flight. They repeat this four times, their white wing-bars electrifying the running black clouds.

A lawn, when you come to think of it, is nothing but a meadow in captivity. When the British moved off the land in the nineteenth century to work in factories and towns, they could not quite bear to leave their rural roots and so created a patch of familiar green behind the house.

Alas, modern lawns have little wildlife value. Most are green deserts, marinated in chemicals, comprised of only a couple of grass species, and shorn stupid once a week in summer. But in the Middle Ages, a lawn was more like a meadow; it was a 'flowery mead', bursting with perfumed wildflowers and herbs and grasses.

These gorgeous, semi-wild acres were an integral part of medieval life, used to their full for walking in, dancing on, sitting in. And in houses and castles where privacy was hard to find, they were the perfect places for lovers to share secluded passion:

He had made very beautifully a soft bed out of

> the flowers. Anybody who comes by there knowingly may smile to himself for by the upset roses he may see tandaradei! where my head lay.
>
> If anyone were to know how he lay with me (may God forbid it), I'd feel such shame. What we did together may no one ever know except us two one small bird excepted tandaradei! and it can keep a secret.
>
> Walther von der Vogelweide (c.1170–1230)

In a flowery mead flowers had symbolism, as well as beauty. Cowslip was Our Lady's keys; daisy was the emblem of purity; forget-me-not was Our Lady's eyes; foxglove stood for Our Lady's gloves.

This I have to confess: there is nothing beautiful about Lower Meadow at the moment. The first flowers have not yet appeared, and the grass is thin and vapid, except for some rank wheaten tussocks where the horses dunged in the autumn. Even sheep will turn their noses up at the grass around a horse latrine. The field is a minor, lower-scale note of green. If that. And stubbly-short, and corrupted by the mud from the sheep's cloven hooves, and by molehills and piles of horse shit which the weather has inexplicably failed to break and level down. The sunlight is spotlight cruel on the grass. Each blade is clearly discernible in its earth surround; a hair in the follicle of the planet. Few blades are longer than 3.25cm, and they are identical.

In this season the grass species cannot really be told the one from the other. But why do a peculiar few blades catch the microscopic breezes and tremble – and others do not?

It takes a moment of effort to see the tiny doily-carte leaves of incipient buttercups, the miniature clubs of clover, the mini shields of docks and nettles. Across the valley are fields that are already fluorescent green and table-top smooth. These are the fields doused in nitrogen. Say what you like about artificial fertilizer: you do get a nice shade of green.

I spend the morning, in the euphemism beloved of horse-owners, 'poo-picking' , shovelling the excess of horse excrement into the link box on the back of the tractor to dump on the manure heap.

A scattering of jackdaws plays a game of its own devising in the turquoise sky. Small spears of lords and ladies have pushed through the earth into daylight.

15 FEBRUARY Dawn chorus at 6.45am. Thrushes going full steam; in the background at Little Trelandon, jackdaws.

Then it rains with high wind. A sweet fierce rain song. The field is desolation, not a beast or a bird to be seen. The Escley roars in the night.

17 FEBRUARY The Escley is quieter now; on the bank under the thicket the tunnels of the water voles have been cruelly exposed by the lately rampaging water. The river bottom, in long stretches, has been wire-brushed of all the silt, algae and detritus down to shining pink-and-green bedrock. Breathing in, I can almost smell pure, exhilarating oxygen coursing above the water.

Beside the copse an oblivious male blackbird tosses aside leathery hazel leaves looking for morsels.

A solitary male pheasant crows in the alders of Quarry Wood. The high dome of the evening sky is wiped free of cloud; a rose glow settles on the spreading girth of Merlin's Hill.

19 FEBRUARY On the lane at night driving home from Freda's parents' evening, the headlights of the car pick out scatterings of pale leaves. Except they are not leaves; they are the white throats of hundreds of silent gazing frogs. The track down to the farm is similarly littered with befuddled amphibians. Penny picks a cautious way around the potholes and the frogs; the four-hundred-yard track takes five minutes for the car to descend. Even so, it is impossible not to crush some of the *Rana temporaria*.

The frogs that reach the house still have three

hundred yards to go to the ditch-cum-pond in Lower Meadow where they habitually breed.

22 FEBRUARY The ditch in Lower Meadow sounds like a steelworks in the evening, so loud is the noise of mating frogs.

23 FEBRUARY The toads are also on the move to their ancestral spawning grounds.

In 'Some Thoughts on the Common Toad' George Orwell posited the emergence of *Bufo bufo*, frog's cousin the toad, from its winter hibernation, after some awakening shudder in the earth or rise in temperature, as the most appealing sign of the coming of spring. The toad, he noted, unlike the skylark and the primrose, 'has never had much of a boost from poets' and added that fasting gave the toad 'a very spiritual look, like a strict Anglo-Catholic towards the end of Lent'. Whether it's high-mindedness or light-headedness, toads, as they crawl to their spawning grounds, are more maddeningly careless of their lives than frogs.

In the morning, there is the squashed body of a warty toad every yard on the lane. But away from cars toads have a good chance of life; they emit a poison, bufotoxin, from their skin to deter predators.

A share of the toad survivors heads with unstoppable, will-not-be-diverted fervour for the shallow flood ponds at the bottom of Bank Field. Haughtily, they do not share the ditch in Lower Meadow with the frogs and newts.

With some squeamish caution I go down at night to the grassy watery depressions. I turn on the torch; there are roaming gangs of toads everywhere. There is nothing pleasant about toad sex: the males are wantonly copulating with each other, five or more fighting for one female in a so-called 'mating-ball', and a dozen are straddling a stone. I push a stick into the water; one toad lurches at it, embraces it and does not let go when I raise the stick in the air.

Next morning the bodies of two female toads are lifeless in the water. They have been drowned in the mating frenzy.

More successful couplings have produced jelly-coated strings of black eggs.

An underrated writer on nature, George Orwell. It is there in the books, if you care to look, such as the detail on woodland plants in *Coming Up for Air*. He was surely right on the democracy of spring:

The point is that the pleasures of spring are available to everybody, and cost nothing. Even in the most sordid street the coming of spring will register itself by some sign or other, if it is only a brighter blue between the chimney pots or the vivid green of an elder sprouting on a blitzed site.

Fields are not natural, although they have natural precursors, the glades of the forest and the grassy uplands. They are man-made with wild materials.

How to make a meadow: the first fields were hacked from woods with stone axes, trees burned out or ring-barked to die a lingering death. Neolithic farmers in this oak-covered valley, as elsewhere, tended to first make meadows on high ground (where the wildwood that had covered Britain in the wake of the retreating ice sheets at the end of the Ice Age *c.* 10,000 BC was lightest) or from the glades that erupted where trees fell down. These Stone Age farmers, though, were gypsies, moving their primitive cattle and sheep on to pastures new as soon as grazing was exhausted.

The trees have not given up; they are still trying to re-forest by sending up blackthorn and hawthorn shoots into the grass, or dropping pips and nuts as fifth columns.

Little in this marginal valley changed for twenty

thousand years, although the Romans, who just succeeded in bringing it within their pale with a fort in the village of Longtown, may have tackled the alders of the trident parallel rivers, the Olchon, the Monnow and the Escley. In 1086 the valley made its bow in history, with a brief entry in the Domesday Book: 'Roger of Lacy also has one land called Longtown within the boundary of Ewias. The land does not belong to the castellery nor to the Hundred. From this land Roger has 11 sesters of honey, 15 pigs when men are there and [administers] justice over them.'

In other words, pigs running in oak woodland was the main farming enterprise in the valley.

To date the field, or at least its beginnings, is not difficult. By Hooper's rule the hedge dividing the field from the Grove farm next door is 600 years old; which is the same date as the thicket along the river bank. For two hundred years the field was part of a larger space, until it was divided again; the Marsh Field hedge is 350 years old. The enigma is the scrawny northern hedge, which is just 100 years old.

Animals made the fields, as much as men: the sheep and oxen of medieval farmers prevented the fields from re-establishing themselves as forests. And nourished the soil with their excrement.

24 FEBRUARY The first clumps of frogspawn are laid in the 'newt ditch', as we call the demi-pond in Lower Meadow.

Few will hatch and survive. The site is known to the heron, fox and other predators.

The luckless day for frogspawn soon arrives. Some time in the night of the 27th a fox pulls clumps of the speckled wallpaper paste that is frogspawn out on to the bank. Yet there are so many clumps of spawn – thirty-four of them, which have congealed into one continental mass – that some frogs will emerge from the predations.

The meadow seems stuck in winter stasis. But then an empty meadow is always in a state of waiting, of anticipation. A wood gets on with things by itself, naturally.

MARCH

Badger

5 MARCH All the birds have burst out singing. A wood-pecker is drumming the dead elm in the valley bottom, as if it has been wound up on elastic, and let go. A sky-lark flutters up over the meadow in its first territorial song of the year, a kite on an invisible thread.

W

6 MARCH A buzzard flies over me with a twig in its mouth.

More snow, which comes in stinging and horizontal from the north to lie in streaks on the red heather hair of the mountain. Even under the hedges, only the dog's mercury can be bothered to flower (pallidly, at that) and the field returns to its winter sleep.

Not quite. The badger has been digging up the dog's mercury roots; it has also tried to break into the top of the warren, seeking the kittens in its hunger.

At the close of the day I stand and listen to the wind scurrying over the surface of the white field.

W

9 MARCH The first primrose blossoms, through the dregs of snow, to sit enchanting by the Grove ditch, a guiding beacon for the sun.

Primula vulgaris is the first plant to flower in the sward proper. I know what today felt like. It felt like the first day of spring. I can smell it. Under the hedges the ground elder and nettles are coming up.

But there are more reliable guides to the advent of spring than my nose. Such as moles:

Anthropomorphizing moles is an ancient meme. Out riding at Hampton Court, the staunchly Protestant William III was thrown from his horse when it stumbled over a molehill; the king broke his collar bone and died three weeks later, in February 1702. The delighted pro-Catholic Jacobites raised a glass to 'The Little Gentleman in Black Velvet'. Kenneth Grahame with *Wind in the Willows* and Alison Uttley with *Moldy Warp, the Mole* are only two of the children's authors to have found *Talpa europaea* irresistible to humanize. Uttley's Moldy Warp took his name from *moldwerp*, the Old Saxon word for the animal, which means earth mover; and moles do shift earth, about 10kg every hour. The spoil is pushed up vertical shafts, to make the familiar mole 'heaves' or heaps. (Not properly vertical: they come out of the earth at 45 degrees, as do colliers' drift mines.) Moles work in four-hour shifts of Stakhanovite endeavour. See, I too have fallen into anthropomorphism.

What storybooks fail to convey is the violent greed of the mole, which scuttles along its tunnels eating the worms, bugs and grubs that fall haplessly in. There is nothing cute about a mole tunnel. It is a vast pipeline trap. And for a gentleman dressed in a velvet smoking jacket, mole is the most violent diner; he bites off the worm's head, then with his claws squeezes out any earth left in the worm, before sucking it down like spaghetti. Any surplus of worms mole stores in a larder. The worms, beheaded and additionally incapacitated by a toxin in mole drool, are alive but unable to move.

In this field, two years ago, I dug out a hole for a new gatepost. As the steel spade cleaved down through the red clay – so dense that the sides of the hole glistened smooth – it unveiled a chamber of horrors about two feet down: a mole pantry with hundreds of entwined worms. The weather had long been dry, so I presume the meat larder was an insurance against worms going too far down into the earth for the mole to find.

Some of the worms were in that state of incapacity known as death. You think of a field as a place of life and growth. And so it is. But it is also a giant graveyard. I have buried dead sheep in the field, as I know, from digging holes for fence posts, others have before me. And then there are all the wild animals that die in their burrows.

This borderland, and quite likely this meadow, saw thousands of years of skirmishing between invaders and settlers, between law-upholders and rustlers. (You need a bit of space for a decent scrap, thus the *field* of battle.) The Saxons, initially, gave up colonization six miles to the east; but burned the halls and hovels of Longtown in 743 then pushed beyond Offa's Dyke in the tenth century. The Vikings harried the valley in 915; the Vikings and the Welsh under Gruffydd ap Llywelyn, Prince of North Wales, rampaged through in 1055; it took the Normans three years to subdue the area's Mercian king, Edric; the Normans in turn gave the unquiet land to Walter de Lacy, who gave his name to Ewyas Lacy, later to be called Longtown after its distinctive one-street style of settlement. The castle built by de Lacy and his heirs was one of ninety the so-called marcher lords constructed to keep the borderland in check; not that they were entirely successful, for the valley became synonymous with cattle and sheep rustling by the Welsh.

Remembrance of those times lingers in the folk mind. Anything well built, from the studded oak door at Clodock church to a well-strung fence, still earns the accolade 'That will keep the Welsh out.' By law, it is still permissible to shoot Welshmen in the cathedral precincts of Hereford with a bow and arrow.

The violence of humans lessened after the centuries of the thieving Welsh, as the border was brought within the proper ambit of England. But bloodshed did not wholly leave the land. During the Civil War a Scots Parliamentarian army camped locally before besieging Hereford. (The Parrys, unusually for a Herefordshire family, were solid for Cromwell; one of them became a colonel of horse in the New Model Army.)

There are days in a desolate November when you still hear the hollering of fighting men, of horses' hooves pounding on the shingle of the Escley. And where are the dead men buried? In this brookside field, probably, where the clay is relatively easy to dig into, and the impenetrable sandstone is deeper than the bottom of a grave.

The gentle pasture of England is tomb after tomb of animals and man, roofed with green.

In this blood-red earth the little miner today is going about his business with gusto. About a quarter of an acre on the upper side is splattered with heaves, and I am reminded of the poet Cowper's line on a mole infestation in a field, where

at ev'ry step
Our foot half sunk in hillocks green and soft
Raised by the mole.

By going on tippity-toes I get to within ten feet of where a mole is digging, volcanically spewing up soil from the centre of a mound. Once, when I was small, I lay with my ear over a mole run in my grandparents' orchard; a small lifetime of waiting ear-to-the-ground was rewarded by the sound of scuttling claws and a distant squeak. Today I momentarily see those claws as they thrust a load of soil up through the hole; they are outsize, splayed, human in their pink nakedness, but with nails from a Halloween witch. A twitchy, fleshy snout follows them into daylight. The mole is probably a male. Male moles suffer a kind of OCD where they make straight galleries. This one is throwing up mounds that run a ruler-straight line towards the centre of the field. There is a great deal of method in his madness; he is digging parallel to where the ditch-water leaks into the field. He is digging exactly where the ground is softly malleable yet not so wet his tunnels will flood. Since March is the beginning of the breeding season he is tunnelling so frenziedly because he is searching for a mate. Moles breed between March and May, when the sows make oestrus to attract boars, the pheromones being switched on by either the lengthening of daylight or the warming of the earth. Gestation is forty-two days, with between three and six hairless pups born in a chamber lined with grass.

After mating, boars go on the hunt for other,

unmated females. If they encounter males in their tunnels they fight, gladiatorially. So there is violence in a field below and above ground.

If I am honest, I am mole-watching because it is easier than my intended purpose in the field, which is a labour Sisyphus would moan at. I am trying to rake some of the heaves flat, because at hay-cutting time they will get chopped into the grass and contaminate it.

I could of course call the mole-catcher, but I quite like moles, and am convinced they do a worthy job of drainage and aeration. Less nobly, I suspect that if you get rid of one mole another will merely take over his or her territory.

Since the banning of strychnine as a mole-killer in 2006, the old-fashioned mole-catcher has made a comeback. One has called here; his gleaming plain-clothes black Merc van was not sufficient to disguise the Shakespearian rusticity of his calling; he killed moles, he told me in a Welsh Valleys accent, with a shillelagh. The dried skins went for 50p to fly fishermen for ties. So, not quite the trade of the 1920s, when mole-catchers wore moleskin breeches and twelve million moleskins a year were shipped to the USA for high-end clothing.

What I do not tell the mole-catcher in his hitman van is that I have always got mole-catchers and the child-catcher in *Chitty Chitty Bang Bang* hopelessly conflated in my head. Or that the mere mention of his trade summons up the four most terrifying lines of English pastoral poetry:

While I see the little mouldywharps hang sweeing to the wind
On the only aged willow that in all the field remains
And nature hides her face where they're sweeing in their chains
And in a silent murmuring complains

In 'Remembrances', John Clare directly likened the cruelty of catching moles in gin traps and displaying their corpses on trees to the pain inflicted on the English rural poor by the enclosure of the fields.

And I know where my sympathies lie.

Anyway, what need have we for mole-catchers? We have Snoopy the Jack Russell. Snoopy has trained the Labradors and the Border Terrier to dig out moles. Their digging, of course, causes more damage to the sward than the moles. Sometimes the dogs bring the dead moles to the house as gifts, little parcels of muscle wrapped in black velvet.

I have returned to the razing of the molehills. As I approach the field, two Canada geese fly overhead, their fruity call confusing the raven in her far fir tree, who calls back. On the field, oblivious in the drizzle, is a young buzzard, one of last summer's hatch. How the mighty have fallen. He or she is eating earthworms, and I do not think it a human presumption to suggest that the buzzard looks none too pleased to be scavenging for so lowly a form of life – the same diet indeed as the mole. Gobble, gobble, quick run to the next worm. Gobble, gobble. Look around. Run. Gobble, gobble. As a style of dining it is more turkey than raptor. Yet appearances are deceptive; inside the buzzard's speckled chest, as streaked and dashed as a thrush's, is a gut, by bird-of-prey standard, of quite unusual length. This means that *Buteo buteo* can extract maximum nutrition from a meagre diet such as the earthworm.

They say that buzzards share with lapwings the trick of jiving on fields so their footsteps sound like raindrops. And duped earthworms come to the surface to be eaten by 'the dancing hawk'. All I know is that two other buzzards fly out of the quarry and stoop to eat worms.

The collective noun for buzzards is, fittingly I feel, 'a wake'.

13 MARCH The frogspawn in the ditch starts to hatch into tadpoles or 'polliwogs'. The latter name for the

larval stage of the frog is derived from the Middle English *polwygle*, made up of the same *pol*, 'head' and *wiglen*, 'to wiggle' – an accurate enough descriptor for these spermy beings.

Like the sea, the red soil of Herefordshire is never constant in its colour. In the banks of the ditch, at the dry top under the lea of the hedge, it is the pink of poached salmon; in the rose haze of another springy, uplifting March afternoon the effervescent mole heaps are the purple of berries, as though the earth is in fruit.

Despite the clenching cold the temperature must be edging towards the 6 degrees centigrade that grass needs to 'flush', to grow with uncontrollable abandon.

Some of the signs of spring are negatives, less about what is a-coming in, more about what is a-going out. The fieldfares are heading north in the evening, the birds, as John Clare declared:

> *That come and go on winter's chilling wing*
> *And seem to share no sympathy with Spring.*

From now until October the sward of the field will never be constant in its colour. Another belting of snow on 11 March is not enough to hide the emergent celestial wood anemones in the copse, or three plucky dandelions in the field proper. Three days later a red-tailed bumblebee (*Bombus pratorum*) flies grass-top height, winding in and out of the fast-multiplying dandelions. I run after it, and see it disappear into a mouse hole in the base of the Marsh Field hedge, down between the leaf litter, the snags of black wool, the upright green of the dog's mercury. The hairiness of the red-tailed bees acts as a fur coat, enabling them to fly when other winged insects cannot bear the chill.

Blackthorn blossom is tight on the branches ready to burst; the thorns that wound round Christ's head are still brutally visible.

For nearly a week I do not visit Lower Meadow, although it is only four hundred yards from the house. My horizon has been reduced to the paddock in front of the house, where the sixty ewes have been gathered for lambing.

Lambing only comes in two ways; either swimmingly well, or drowningly bad, and this year I have spent more time with my hands inside ewes' wombs than either the sheep or I are comfortable with. I have made L-shaped shelters of straw bales or corrugated iron for the ewes to have cover from the driving rain. I like the shelter too when, befuddled at five in the morning (by experience a sheep's favourite time to lamb), I play the live-or-die game, of trying to sort out, inside a ewe, which bony leglet belongs to which lamb. De-tangled, repositioned, the Ryeland lambs emerge in yellow slimy pods, to be rubbed with straw or rags. When they do not raise their heads and bleat . . . a blurred flurry of rubbing, air blown up noses, red stomach tubes delivering colostrum, purple spray on hanging navels.

Our sheep have names, the names a rough-hand mnemonic to colour, type, date of birth, or an imagined resemblance to a character, real or fictional, or to a personality trait: Chocolate, Sooty, Soo, Tiddlywink, Shortbread, Cardigan, Jumper (of course), Valentine, Tess . . . You come to live with us, you get a name.

One runt lamb is born dead, another dies within hours, and for both I grieve with clenched eyes for the life never lived. There is nothing so innocent as a new-born lamb; the scion of the sheep was not appropriated as the Christian symbol for Jesus for nothing. The lamb of God.

All the Shetland and Hebridean ewes birth with a feral ease, the curly black lambs 'sharp' and walking within minutes. No, the wearing habit of primitive sheep is not their birthing technique; it is the flightiness of the ewes when mothers.

Something comes into the paddock at night, sending the ewes and lambs into a baaing delirium. Within minutes I am out with a torch. The intruder has gone. One Hebridean ewe, who has just lambed, runs off, leaving her twins behind; over the course of a greasy grey day I try to reconnect her with her offspring, finally resorting to catching her and penning her so tight she cannot turn, and then put the lambs in to suckle. All she does is jump up and down on them; before they are murdered I let her go, and put them on the bottle.

They live in the sitting room in a dog-crate. Sheep tamed by being bottle fed are no bad thing, since it means they will, when grown, come to food. And the rest of the flock will follow.

21 MARCH Heavy rain. The horses in House Field stand back to the rain, the sheep and their lambs are either under the hedges or tight against the bales. The red-tailed bumblebee must be glad of the house that it has taken from the mouse. In Lower Meadow I see a small flock of forlorn redwings, the thrush with the fetching cream eye-stripe and orange flanks, in the hazel. At my approach, up into the air they go, slipping left, slipping right, drunkenly unsteady.

They loiter for a day. On the 23rd I hear redwings 'zeeping' in the starred night when I'm checking the sheep. Next day there are no redwings on the farm. They have gone north, to home in Scandinavia.

W

With lambing done, we visit my sister-in-law in London for a night. On the return home I am exhilarated by how fresh and enervating the air is, how lovely is the taste of spring water compared to chlorine.

Then there are the benisons of the earth. In the uplifting sunshine I pick dandelions for dandelion wine, the cottage wine. They are the flower of the sun and their faces follow its course. They are nuclear furnaces in miniature.

Less given to whimsy, the French persist in calling dandelion *pissenlit* in honour of its diuretic capability. To me as a child they were 'clocks', whose seed heads were blown on to tell the time and tell the future. And in a sort of fashion dandelions do tell the time: like the white, fragile wood anemones, they close their heads at night.

Dandelions have not always been weeds. In the Victorian era they were cultivated in walled gardens and eaten by the aristocracy in dainty sandwiches.

Then there is the silence of the night, when the dandelions are closed up.

A dog barks somewhere down the valley, another takes up the call, then another. An aural chain reaction of barks proceeds from Clodock to Michaelchurch Escley. Presumably they have all been watching *101 Dalmatians*.

W

25 MARCH The first blackthorn blossom unfurls into delicate white crystals in the hedges around the field. But even such a picturesque setting for the field cannot distract from the drizzle that is centre stage. It is the sort of drizzle that seeps into everywhere, and everything. For an hour I sit under the hedge with constantly wiped binoculars watching a male wren making a nest in Marsh Field hedge, flying in with stalks of grass, doing a distinctive cocky tilt of the tail as he lands on the willow branch. Out he flutters and off. Although the hedges are not in leaf, far from it, the trapped leaves and twigs of yesteryear and of hedge-cutting give cover to his construction. He may well build other nests, which he will display to any female who enters his territory. If she likes any of his pads she will move in, decorate, and bear his children. A slapper seeking a Premier League husband could not be more shallow.

Mind you, he is no moral giant. As soon as he has ensconced one female, he will try to tempt another Jenny Wren into one of his spare nests, where she too will give birth to his progeny. The little cock then travels between his families, a bigamous commercial traveller in a 1930s thriller.

'Wren' has its origins in the Anglo-Saxon word *wrœnno* meaning lascivious; the Anglo-Saxon is linguistically kin to the Danish *Vrensk*, meaning uncastrated. Slightly less fun is the Latin tag for the bird, which is *Troglodytes*. Typically this is translated as cave dweller, although hole-plunger (*trogle* being hole, *duo* to plunge in) would be a more tellingly accurate description of the furtive wren.

Curious how such a randy, though otherwise inoffensive little bird became an object of loathing. In parts of Ireland the tradition of Hunting the Wren is still played out. Groups of boys used to go out into the countryside to capture or kill a wren, which was then paraded around the December village with the youth chanting:

The wran, the wran, the king of all birds,
St Stephen's Day was caught in the furze;
Come, give us a bumper, or give us a cake,
Or give us a copper, for Charity's sake.

Today an effigy of *Troglodytes troglodytes* is used.

This victimization of the wren seems to stem from the bird's role in alerting guards to the attempted escape of the English Christian St Stephen from prison. Other folkloric tales about the bird's indiscreet warnings abound; in the seventeenth century it is said to have hopped up and down on a drum and warned Cromwell of a sneak attack by the Irish. The percussive sound of its alarm call gives rise to the bird's Devon name of crackil or crackadee.

The wonder is that a bird so tiny can make so much noise. Scientifically speaking, birds have no larynx, having instead an organ known as a syrinx. The syrinx is far more efficacious at producing sound than our own larynx and its vocal cords. Whereas a syrinx is able to vibrate almost all of the air coming out of a bird's lungs, human vocal cords utilize a mere 2 per cent of the air passing over them. Additionally, the syrinx can produce two different sounds simultaneously (one from each half), which goes a little way to explaining the complexity of birdsong.

Yet listening to the cock wren tune up is not a matter for science. Few other birds can lift the solitude of a damp March day when winter will not go away. On a similar day Wordsworth encountered a sweetly singing wren:

The earth was comfortless, and, touched by faint
Internal breezes, sobbings of the place,
And respirations, from the roofless walls
The shuddering ivy dripped large drops, yet still
So sweetly 'mid the gloom the invisible Bird
Sang to itself, that there I could have made
My dwelling-place, and lived for ever there
To hear such music.

Many wrens establish winter territories, as this one has. The long forceps bill of the wren is designed for picking the minutest of spiders and insects from the litter of woodland floors. (Reposed in death, a wren looks to be all beak.) Damp grassland with hedges is an acceptable substitute for woodland terroir, so while there is one male making des reses by the newt-ditch willow there is another toiling on the river side of the field, where the trees surround the promontory, and where the frost does not always iron-harden the earth.

The males are trilling against each other, despite the rain, and all the wrens of Britain are getting ready to nest, despite the rain. All 17 million of them.

26 MARCH Plants under the Marsh Field hedge: rabbit ears of foxgloves emergent; ground elder; nettles; ground ivy, already with tiny violet flowers; dog's mercury with white flowers; cleavers beginning to lick the trunks of hawthorn and hazel; basal starfishes of thistles; unfurled hoods of lords and ladies. In the dryness of the bottom of the hedge there are little colonies of holes for voles; because the hedge bottom is raised up by the litterfall of the decades, it is high and dry. In the copse and in the field the yellow lesser celandines are starting to shine.

'Curleee. Curleee.' The sound of the wild. I have been listening out for them for days, and they come on this afternoon of the 26th, planing down the wind as they make their plaintive cries to the souls of the departed.

The curlews are home. To my private amusement our curlews are inverse Herefordians, in that they spend the winter holidaying in west Wales; human Herefordians go to west Wales for their summer hols.

I fret eschatologically about the curlews, as though it is their migratory wingbeats that turn the earth, and should they fail to appear we will have entered some ecological end time. But they are home, home to breed. Curlews are not gregarious in the breeding season, and each couple likes a great deal of space. Historically, we have had one pair nesting in the field by the road, and one pair down in the meadow – which is about as far apart as you can get on forty acres.

A male curlew glides around performing a vibrato bubbling trill. Within two days he has secured a mate and then the curlew pair circle round the bottom field, laying claim to their territory. We love the haunting piping of the curlew with fierceness. For us, it is not just the sound of the wild, or of spirits calling to the dead, it is our personal heralding of spring.

Curlews live for five years, and are creatures of habit concerning habitat. The likelihood is that one of the pair in the meadow was born here, as were its parents. They have taken to the air and are crying again. The elongated double note of their call is almost caught in the curlew's name if the stress is placed on the first syllable.

W

Frost. Frost so hard that the grass is as white as fronds at the bottom of the ocean. A low struggling sun makes tall, flat shadows of the riverside trees that stretch almost across the field.

There are tracks in the frost, bright green lanes of animal traffic. The rabbits have advanced cautiously into the field, scraped into the grass, and back to their burrows on the bank.

But again not all of them were quick enough. In this scene of pearlescent perfection lie tufts of greeny-grey rabbit fur. And drops of blood. Walking towards the copse trees where the frost is still coldly bright I pick up the surreptitious pad marks of the fox, the hind foot placed precisely into the print of the fore foot.

My guess is that the dog fox has taken a gift to his vixen, who is now suckling her cubs. I see him that evening trotting into Bank Field, then redoubling his trotting tracks, working his way into the wind, into the dusk. (I do not think I have ever seen a fox walk except when stalking; like Labradors under the age of ten, they run everywhere.)

I barely recognized him. Gone is the gorgeous coat of winter; his moulting fur is threadbare and tatty.

In the newt ditch the water has again frozen. A solitary silver backswimmer (*Notonecta glauca*) is preserved in a crystal sarcophagus. A temporary tomb, yet dead forever. The topmost frogspawn is also encased in the killing ice.

29 MARCH The tadpoles are not finished with trouble. When the ice melts the heron comes plunging his lightning beak into them. But he must find the oozy bottom of the ditch shiftily uncomfortable or the tadpoles unsatisfyingly titchy because he gives up after only a minute or two to flap off on tired wings.

There is something primeval about herons; they trail ancientness behind them. As it sluggishly flies towards the Grove it calls once; a screech from dinosaur times which terrifies an already melancholic sky.

The surviving tadpoles in the newt ditch stick together, safety in numbers, salvation in a heaving unsaintly black mass, save for a handful of brave explorers.

This is a trick lambs are less able to pull off.

Penny and Tris arrived home to say they saw a red kite sitting on a molehill in the top field. Except they then realized it was no molehill; it was a black lamb, and the kite was pulling it apart. The lamb had been running around happily thirty minutes before.

When I stomp up through the bone-cold mist, the kite is still there, tearing with its beak. Only when I get to within thirty feet does the kite launch up; it makes a half-hearted attempt to carry the lamb off but only succeeds in dragging it for a foot or two, before insolently making its way towards the mountain wall.

The black lamb in its curly astrakhan coat has had its guts slit open, pink and exposed. White heat rises from the body in gasps.

I carry away the dead lamb by its gangly back legs. The eyeless head, heavy with skull, swings and contorts in grotesque stringless puppetry.

Beyond the mountain is the fastness of the red kite. Once upon a medieval time the red kite was about as common in London as the sparrow, and more welcome because of its habit of cleaning the capital's streets with its scavenging. Years and years of persecution by keepers on shooting estates beat back the bird to mid-Wales. By the 1950s the red kite population was down to about a hundred pairs. Through protection it has risen in numbers, and expanded its range.

Today I wish the dam wall of the mountain had held the bird back.

W

30 MARCH The month ends in a blaze of clear-skied glory.

There is one perfect, silver-toned night, when the moon provides lighting, when the Escley has settled down to an expectant hush, and there enters from stage right a white ghost which drifts silently across the field.

A barn owl. *Tyto alba.* Its pale flat face gives me an unconcerned glance. Barn owls have peerlessly acute hearing, and the owl is listening for the squeaks of rodents rather than watching for their movement. When the owl reaches the thicket it banks right and makes a return pass over the field. The barn owl is the owl of meadowland: detecting noise with its asymmetric arrangement of ears is easier above grass than woodland, with its rustling interferences. On this night nothing catches the owl's ear, and it veers off to the Grove. Lower Meadow is not the barn owl's usual hunting ground for it prefers the still more open aspect at the top of the farm. This protractedly chilled March is making for desperate hunting measures.

Indisputably, there is something spooky about barn owls. They are the demon owls and death owls of country lore. Shakespeare frequently employed them to dramatic effect, and nowhere better than *King Henry VI*, Part III, Act V, Scene 6, when, at the hour of his murder in the Tower, King Henry tells the villainous Richard of Gloucester, 'The owl shriek'd at thy birth.'

The moon disappears behind a monstrous cloud. On cue, the barn owl emits its territorial cry from somewhere in the darkness of the Grove. Barn owls do not hoot. They screech, they scream. With the anguish of a dying child.

Across the Escley in the wood of the old quarry the tawny owl, the wood owl, the ivy owl, the brown owl, emits its comforting 'tu-whit'.

APRIL

Cuckoo pint

MORDANT CLOUDS FLOOD over the mountain, and the field is wreathed in the dark that comes before the storm.

There is a single bright spot. Today on this 2 April the first cuckoo flower in the field blossomed, to nod its pale pink bloom in the gathering wind.

Any flower that comes with a host of local names is likely to be of human use, either as food or as medicine. The cuckoo flower has at least thirty local names, among them lady's smock, milkmaids, lady's mantle, lady's glove, cuckoo's shoes, which usually point to its habit of flowering to meet the arrival of the cuckoo, or innuendo-ishly, to its passing resemblance to women's undergarments hanging on a washing line. The vernacular meadow bittercress is the most useful of names, since the needle-thin leaves of *Cardamine pratensis* make a peppery edible that used to be sold on medieval market stalls. Left uneaten by humans, cuckoo flower is the foodstuff of the caterpillars of the orange-tip butterfly.

Cardamine pratensis shares one country name, 'cuckoo pint', with the wholly different lords and ladies (*Arum maculatum*). All the folky nomenclatures of lords and ladies are of eye-winking, *Carry On* standard. Cuckoo pint here is derived from cuckold and 'pintle', meaning penis. Geoffrey Grigson lists as many as ninety nudge-nudge alternative names for *Arum maculatum* in his book *An Englishman's Flora*. So: cuckoo cock, dog cocks, kings and queens, parson's billycock, stallions and mares, and wake robin (Robin being the medieval equivalent of 'Dick'). Such names are saucy souvenirs of British rural humour.

The plant's shiny halberd leaves (with disfiguring black poxy spots) have been visible for a month in the hedges, but now the brown phallic spadix is . . . tumescent. It is a flasher in the hedge. On the warm days soon to come, midges, enticed by the meaty smell of the spadix, become trapped in the outer suggestive sheath from which it peeps. The midges will fertilize the hidden flowers, which in turn will become the beguilingly orange berries of autumn. At night the sheath will loosen, allowing the midges to escape.

The tubers used to be made into a love potion; in John Lyly's play *Love's Metamorphosis* of 1601, he had a character say, 'They have eaten so much wake robin, that they cannot sleep for love.' Whatever their efficacy as medieval viagra, the plant's roots, properly prepared and baked, made a kind of arrowroot once sold as Portland Sago, and were the main ingredient in 'saloop' (salep), a working-class drink popular before the introduction of coffee and tea.

Gilbert White recorded thrushes eating the roots in severe snowy seasons, and the berries are devoured by several kinds of birds, particularly by pheasants. No animal will touch the leaves. They emit prussic acid when bruised.

Drops of rain linger, glisten, cling to the sword-blades of grass. They do better than a bronze beetle that climbs a grass stalk, falls, climbs, falls, never achieving escape.

A peacock butterfly is on the wing in the floating spring air, and nectars on the cuckoo flower. The butterfly spreads its wings as it feeds, advertising its garish eye-spots. They watch over the field, but really they are devices to deter predators. They are passable imitations of the eyes of a giant bird, of an avian monster. The pansy-coloured peacock is happily unmolested by birds, even when it provocatively warms itself on a flat stone.

Birds are in and out of the hedges in a constant traffic of nest building. Chaffinch. Great tit. Blue tit. Robin. They still seek the innermost sanctums because the hedges are still not in leaf. A favoured material, I note, is dried stems of grass from the meadow. The nest building of the birds invisibly binds the field with the hedge. Passing sunlight rebounds off the smooth woman's skin of the hazel.

W

I am not the only farmer in the field. Away in the rough grass beside the Grove ditch are three tumps of yellow meadow ants. Estimating the age of anthills is approximate, yet not so vague as to be useless. *Lasius flavus* digs down into the earth and brings up its spoil at the rate of a litre or so a year; it is the spoil which makes the hill. The colonies beside the ditch are about five years of age; the home mounds in Bank Field from which the winged agates (in non-science speak, flying ants) flew on their drifty summer's voyage of colonization are twenty. The steepest part of Bank Field burgeons with so many mounds it appears the earth has boils.

There is no finer soil than the soil of the meadow ants' nest, for each individual particle is dug out and then hauled by the worker ants to take its place on the mound top, all stones and debris left behind. A few select blades of grass grow from the bald earth dome, like hairs on the head of a venerable curate.

Since the mound is above ground it catches the sun; the ants use it as maternity ward and nursery, and will even carry the eggs through the mound's network of tunnels to the warmest side. Unfortunately for the ants, the fineness of the dome's soil and its elevated position make for poor protection against predators. Thin badger sows in spring sometimes break open the great mounds in Bank Field with frantic bear swipes in pursuit of ant larvae, or better still, the eggs.

But it is not a badger that has attacked one of the anthills in the meadow; the damage is too small. The culprit is a green woodpecker, which has stabbed into the anthill with its beak, wrecking about one half of the dome. In the spirit of scientific enquiry (though with a nagging sense of hooliganism) I dig a spade into the ruined earth, a couple of inches at a time, down through chambers and passages. I am too careless to begin with, and have to slow down to the speed of an archaeologist. The ants themselves robotically pick up disturbed eggs, as though a metal blade bisecting their home is an everyday, hey-ho, experience.

Eventually I locate my prize. In a cellar room about the size of a ten-pence piece is a small herd of grey soporific aphids. These are kept captive by the ants, and 'milked' for the sugar or 'honeydew' they excrete. The aphids themselves feed on the roots of plants in the dungeon's roof and walls. This is intensive farming of a manner to make any agri-businessman green with envy because the aphids are selectively bred; in all probability the little herd of aphids in this chamber are clones of a good 'milk aphid', an insect Holstein cow.

Meadow ants are full of surprises. They can live for twenty years, and in dry downland areas they will take the larvae of the chalk hill blue butterfly into their nests and raise them.

Meadow ants are not really yellow; they are the ginger colour of tea made by grandmas.

W

By 12 April the brightness of the flowering celandines means that to cross the meadow in the evening is to walk through a starfield.

The blossoming of the flowers is now unabated; the first bluebells burst out in the copse, and within the week the campion comes out under the hedge, and there is secretive stitchwort there too.

Flies waltz the warming air. The ground temperature is constantly above the 6 degrees Centigrade that grass needs to grow. There is another necessity for the greening of the grass; meadow grasses need, depending on the species, between ten and fifteen hours of daylight for the uprising.

Standing in the middle of the field at night: someone has stirred the clouds into milk pudding.

W

I am sitting on the bank of the existential river. Upstream of the tree-hung pool there is a dipper on an algae-glossed boulder. These members of the thrush family are the riparian equivalents of canaries in a coal mine. In a pure river like the Escley, with its abundance of crustaceans, bugs and fishes, the dipper density is about as high as it gets, and the two hundred metres of river along the meadow support one dipper pair. It is the male on the stone. I know he has seen me, because he is 'dipping' – bobbing up and down to show off his startling white shirtfront. This is a signal from bird to predator that the latter has been seen and has no chance of a sneak attack.

The bird dives into the pool, more gainly than one might suppose from a blackbird lookalike, and pops up with a struggling bullhead. The death of the bullhead is brutal, seized by the tail to have its head bashed out on the green boulder.

When the prey itself is stone-still, the gentle dipper flips it up and swallows it head first. With one beat of its wings, the bird then glides downstream to the shallow water running glassy and cold over the shingle. The bird walks along into the current, peering monomaniacally. A tyrannosaurus dart of the beak, and a caddis fly larva is extracted from the sheeny water. Its bristly catch in its beak, the dipper flies off low around the bend, to its moss-lined nest in the bank, where it will squeeze the caddis fly larva into the mouths of the waiting dipper babes. The dippers have used the same nest in a slit in the sandstone, swaddled by the elm's roots, for the five years we have lived here. For all I know, dippers have been using the nest for decades, even a century or more. They are birds of tradition, with successive generations using the same nest.

W

April: the month of greening, of greenshift, when everything bursts into leaf and growth. Squatting by Bank Field hedge, taking a spirit-level perspective towards the river, it looks as though the floor of the field has risen by two inches. Actually, my eye is not so far out. I have my ruler with me; the grass has grown in the spring flush by an inch a week over the last fortnight. Behind me in Bank Field the ewes and lambs are feasting on the verdancy, the lambs breaking off to play king of the castle on the fallen trunk of the elm, which lies like a tossed-away dog bone, and which nobody in thirty years has got around to moving. Such are the unintentional conservational benefits of laziness that the prone elm hosts beetle colonies galore; the foxes have been digging them out, and the wing cases (elytra) in their scat dumped on the tussock by the gate catch the last shards of sun.

After a while the unknowing lambs in their evening gangs realize they have become separated from their mothers, and start up with plaintive calling. All down the valley lambs take up the mayday, so it reverberates around the hills.

The Victorian naturalist W. H. Hudson would spend a whole day in spring just admiring grass, 'to rejoice in it again, after the long wintry months, nourishing my mind on it . . . The sight of it was all I wanted.'

At twenty-four inches in length, with a preposterously long down-curved bill, the curlew is an outsize and distinctive wader. Put it down in a field, though, and it disappears in a Houdini piece of legerdemain. It takes several sweeps with the binoculars until I locate the female, who is pulling at a clump of dry grass. The male has already scraped a depression in tall sward about twenty yards out from the hedge; his DIY has been done half-heartedly in a manner a man would understand and a wife condemn.

Two days later she is sitting tight on her eggs. To help me locate the nest again I tie a white rag in the hedge directly behind it.

I have taken up observational residence in the bottom of the far hedge, the four-foot isosceles triangle where the hazel has broken down and nettles rampage and the sheep shelter from the sun. Every field should have a neglected corner. While I can peer through the shambles of decaying hazel across the field and see almost everywhere, it is sitting inside this gone-feral space that I am most aware of the immediacy of beauty, the beauty of immediacy. The hazel screen obliges concentration on the things that are close. There is the cough-mixture whiff of ground ivy, and the whirling black flies so small that I can barely see them, whose name I do not know and never will. The vine of the ivy winds up in a faultless helix. Then I see the paradise blue hue of the dog violet ('dog' being an unkind reference to its lack of perfume), the pale-green towers of Jack-by-the-hedge, which might be better named as Jack-beanstalk. But rub a leaf, and you will know why it is also 'garlic mustard'. Have you ever stopped to notice how perfect are the curves of an earwig's rear pincer? Or how like amber an earwig's body is?

I am transfixed by my own prison; through the bars of the branches, however, and past the skittering light I am not oblivious to the fox, because movement always gives the predator away, as surely as it gives the game away. The fox knows the curlew is there somewhere in the field. It stands intent, it sniffs and it stares. The curlew does not move. Curlew make good eating, and used to be as popular on the human table as in the fox's den. According to poulterers' prices fixed by order of Edward I in 1275, the curlew was 3d a curved head.

Neither by sight nor by nose does the fox locate the curlew. And it lopes away, disgusted at its failure.

16 APRIL Note on piece of paper put in pocket on walk around the field with the dogs in the morning: 'More primroses out on south side; blackcap singing, & the chiffchaff too.' These are the first of the summer migrant warblers to reach the field. The chiffchaff does not stay, and moves on. The blackcap sings from the top of Bank hedge, and I cannot help but have my heart stumble in admiration. The complexity of the blackcap's song was pinned perfectly by the French composer (and ornithologist) Olivier Messiaen, who used it as the musical symbol for the eponymous *Saint François d'Assise*. 'I had to insert,' wrote Messiaen, 'chords of each note in order to translate the special timbre, which is very joyous and rich in harmonies.'

The blackcap deserves its title as the 'northern nightingale'. Except for this: the bird's alarm call is a crude 'tak', as though two pebbles were being chipped together. For the entire summer the blackcap 'taks' at me, at the sheep, at everything with the constancy of a dripping tap.

Of all the summer martins, the swallow is the one which spends most time hawking the field, and on the

20th I see the first ruby-face of the year. Swifts and house martins all take their turn over the field, but they feed higher; it is the dearth of high-flying insects outside high summer that accounts for the shortness of the swift's sojourn in Britain. Today's rain has forced the insects low, and the first swallow of summer is doing what swallows do best; lacily, elegantly skimming over the sward-top after winged prey. (Male swallows with the longest tail streamers, incidentally, are the most attractive to the girls.)

The delicacy of the moment is ruined by the two Canada geese, who come honking over to land in the lake further up the valley. Vulgar, with absolutely no subtlety, they are irate drivers in an LA traffic jam.

The French call the yellow wagtail the 'shepherdess', and true enough the bird does follow on behind the sheep, hoping for insects turned up by ovine cloven hooves. It is much happier in Marsh Field, where there are not only sheep but acres of soft ground. A pair nest there among the sedge, and sometimes flit over into the meadow to explore the wet corner. The male enjoys perching on the hedge and singing. If it can be called singing. He may have the yellow plumage of the canary, he does not have its voice. All he can manage is an insistent 'te-seep'.

They run as daintily as fairies, and the male is a tantalizing flash of gold in the grass as he hunts about. They arrived on a raw Friday at the end of March, but they bring warmth to every place they tread, hence their old local name of 'the sunshine bird'.

Almost no birds today have vernacular names. Bird names have become standardized, homogenized, conscripted into what is considered proper by scientists for classification. A century ago a birder could have told what county, even what village, he was in by the folk name for a long-tailed tit. In his *Treatise on the Birds of Gloucestershire*, W. L. Mellersh collected no fewer than ten local names for *Aegithalos caudatus*, the long-tailed tit, among them long Tom, oven-bird, poke-pudding, creak-mouse, barrel Tom, and in the south of the county, long farmer. For John Clare in Northamptonshire the long-tailed tit was, delightfully, 'the bumbarrel'.

18 APRIL Bluebells out in force in the copse, making a blue gas mist over its floor, an uninterrupted mat of docks, celandines, wood anemones (alas, on the fade).

The meadow pipit launches off a fence post, and ascends flutteringly up to twenty metres, till it reaches damn near the top of the young oak, accelerating its 'sweet-sweet-sweet' song. Then it falls anxiously back down on half-spread wings, with a valedictory and tuneless trill. It's an apology for birdsong against the neighbouring skylark's joyful riot. But I sympathize: I can't whistle a note either.

And then the mist descends to put a slate lid on the valley and its proceedings.

W

St George's mushroom (*Calocybe gambosa*) is one of the earliest mushrooms to appear, traditionally making its creamy bow in the green grass on 23 April, the day commemorating England's patron saint. Due to global warming it has cropped earlier and earlier, but the long cold of this spring has encouraged it to keep close to its historic calendar date. I notice a 'fairy ring' of the mushroom on 22 April in the usual place, about twenty feet in from the north, Bank Field, hedge. With its convex head and well-proportioned stem, all in Classical order, St George's mushroom is handsome rather than pretty; it looks good enough to eat, and is. And its aroma, my nose to ground like a truffle hound, is alluring too. The mushroom smells of flour.

W

There was an unexpected visitor in the field today. As I walked down the bank in the morning haze the blackbirds were clamouring their liquid alarm. Then: dismissive wasp-yellow eyes. Scaly yellow legs. Black metal talons. All these things flashed before me.

I am not sure who was the more surprised, the female sparrowhawk or I as she came up over the hedge. I could feel the displaced breath from her wings as she flicked up over my head, then away, a sullen grey bullet.

Certainly I was the more scared; for malevolent verve the sparrowhawk is unrivalled. They are always coiled, ready, dangerous. When the first gunsmiths needed a name for a small firearm they settled on the falconry term for a male sparrowhawk. A musket. But if anything, the female of the species is more deadly still; at ten centimetres bigger than the musket she can take a speeding wood pigeon out of the sky.

Maybe five times a year I see a sparrowhawk on the farm, usually in summer, when they dash after an ascending skylark or meadow pipit, so beautifully but foolishly advertising their presence. Today the sparrowhawk has hunted low and swift around the hedges, and burst in among the chaffinches; there lie the remains of the bird on the grass in a crown of plucked feathers. Sparrowhawks sit on their grounded prey, so their talons pierce the body, and if this is not enough to administer murder they make darting stabs to the back of the neck. The meal-bird is defeathered at the top of the chest, just as humans are shaved for surgery; into this bare flesh the sparrowhawk inserts its bill to begin its gluttonous surgery.

Sometimes a kestrel hunts the field, sometimes the red kite, and I once saw a merlin. Of the diurnal raptors, through, the field is truly the hunting ground of the buzzards from the quarry wood. One flies above me now, beating the bounds. His patch. My patch. This field is the space we share.

22 APRIL There are now so many cuckoo flowers that the boggy corner looks like a city of lights. The arriving swallows no longer hunt a green sea, but are now skimming over a meadow of flowers galore. The field forget-me-not, with its startling yellow eye framed by blue, has also debuted. The wildflower days are here.

Night in the field. On the far horizon to the south there is an unsightly smear of urban light. Otherwise the night is the black of deep space, the original black of the universe.

A pair of car lights come along the lane which runs

along the spine of the hill. Cars, though, are still sufficiently uncommon to be romantic, as though the people inside were on their way to some secret assignation. Then comes the milk lorry, punctual but out of time. One by one the dairy farms in the valley have closed down. What price milk production at a penny a litre profit even on 'intensive' grass and Frankenstein cows with overdeveloped udders?

The night returns to its perfect pitch. A rabbit, across in the old quarry, pig-squeals as jaws clamp it. The foxes or badgers are about. I settle further down into the depths of the hedge.

In this state of blindness, hearing becomes more acute. (Later I can even hear the tawny owl making its hunting flight.) From the field comes the slightest scuffling. I turn on the torch, and there it is. A mouldywarp. A mole pup, wallowing through the grass and floral waves.

When mole young are five weeks old they are ejected from the nest by the sow to make their homes in nearby tunnels, either by taking over existing runways or by excavating their own. Soon the sow tires even of this proximity to her offspring, and there is a second diaspora at the end of the summer, when the mole young will take up home hundreds of yards from the maternal burrow. The dispersals are done overland, and leave the moles vulnerable to predation. Predators abound, and bound. Mole pups are a favourite of owls, foxes, badgers, weasels, stoats and the polecat that lives down the lane.

It is not true that moles are blind; the light from my torch startles the mouldywarp, which stops and sniffs with its snout. I turn off the torch and let the mole go on its way. Black into black.

I must be this mole's guardian angel. When I switched on the torch it picked up two amber eyes only yards behind the mole. The vixen. She turned and walked off, haughtily aware of her power. She only has to be lucky once; the mole has to be lucky always.

All the little mouldywarps sail forth on the following nights. The odd thing is that I never discover the fortress, the super-tump under which the sow has her nest lined with grass and leaves, which must be somewhere in the depth of Marsh Field hedge. Not all the field's secrets will be given to me, it seems.

24 APRIL There is something intensely uplifting in seeing the house martin, who twice a year undertakes a dangerous migratory journey to build his house here, as though this place was perfect.

Shakespeare too had a particular liking for the 'martlet', which he identified as a symbol of beneficence:

This guest of summer
The temple-haunting martlet, doth approve
By his loved mansionry, that the heaven's breath
Smells wooingly here: no jutty, frieze,
Buttress, nor coign of vantage, but this bird
Hath made his pendant bed and procreant cradle.
Where they most breed and haunt, I have observed
The air is delicate.

Birds have a Proustian capacity for making remembrance. I only have to see a house martin and I am in my childhood home, the windows of my bedroom open, head out, watching the chattering, surveying house martins build their intricate mud cups under the white-painted eaves.

April showers? I would settle for April showers, I would settle for anything short of heavy rain in this soaking fag end of the month. Fortunately Thomas Hardy's darkling thrush is made of cheery stuff and sings matins from the top of the elder. Perhaps he knows of better weather. The field is more than sodden; it is inch-deep in water. In such times as these farmers make poor jokes about planting rice in paddies. To remind me of the wateriness of the world, two mergansers land on the river, and when walking up from the field with Freda a strange bird flies at head height past us. 'A flying chicken!' jokes Freda. No, not a chicken; a web-footed great crested grebe, the first I have seen here.

Real seasons with real weather do not progress smoothly. They stop, they start.

MAY

Curlew

MAY IS NAMED for Maia, the Roman goddess of growth. And the increasing heat of the sun does bring on life. The greening suddenly becomes unstoppable, overwhelming, deliciously frightening. By the 3rd the grass in the meadow, in all of a rush, has reached a foot high, and if I lie on my elbows I am floating on a pea-green sea into which someone has thrown a confetti of blooms. Now I too have Hudson's 'spring grass mood'. I let the cows out of their winter paddock, into Marsh Field, only two days after the traditional day for moving cattle on to summer pastures. Quite taken with the mood of the moment, they run around throwing up divots. Dancing cow day we call it, this day when the cattle are released to munch their way through knee-high Maytime flowers.

And the cowslips unfurl their Regency-bonneted heads in the meadow. As flowers they have benefited from a useful historical amnesia; the 'slip' in their name derives from the Old English *cu-sloppe*, meaning cow slop or cow shit. The charming, antique yellow *Primula veris* does indeed grow best in meadows where cows lift their tails.

The air screams. The swifts, on their mechanical

bat wings, vortex around the house until it is time for bed. They arrived yesterday.

W

5 MAY For weeks my ears have been straining for the sound of the cuckoo from Africa. 'Was that a cuckoo?' I say to myself, to everybody, every time I catch a half-bar of a particularly tuneful cooing wood pigeon. But today I do, without doubt, hear a cuckoo, down the valley, while I am swimming on my ocean.

I only hear the cuckoo once. But a century ago, on the hills above Buxton, Hudson found:

> From half-past three they [cuckoos] would call so loudly and persistently and so many together from trees and roofs as to banish sleep from that hour. All day long, all over the moor, cuckoos were cuckooing as they flew hither and thither in their slow aimless manner with rapidly beating wings looking like spiritless hawks.

The resistible decline of the cuckoo has come to this: I hear a solitary cuckoo on a single occasion in a whole valley in a spring. The cuckoo is now on the red list for Birds of Conservation Concern. Welcome to the cuckooless spring.

At least the meadow pipits in Lower Meadow will

be pleased with the demise of the cuckoo. The meadow pipit's nest is often the favoured choice for the cuckoo to lay its Trojan egg. Indeed, so closely associated is the meadow pipit with its role as the unwitting foster parent of the parasitic cuckoo that in Welsh the bird is *Gwas y Gog* (cuckoo's knave).

Meadow pipits are the mugs of the bird world, the victims of the malevolent con artist cuckoo and prey for charismatic merlins, hen harriers and sparrowhawks. Foxes and weasels predate their eggs. But I agree with Hudson that no one who sees the speckled bird 'creeping about among the grass and heather on its pretty little pink legs, and watches its large dark eyes full of shy curiosity as it returns your look and who listens to its tinkling strains . . . as it flies up and up, can fail to love the meadow pipit – the poor little feathered fool'.

There are two meadow pipits' nests in the field, both with four dark brown eggs in their cup of dry grass. These are incubated for thirteen days. In both cases my attention was drawn to the nests by the courteous males bringing food to the sitting hens. Meals were mostly spiders, moths, grubs and caterpillars, almost all hunted within the confines of the field.

Like the cuckoo, the meadow pipit is in decline. Indeed, the national loss of meadow pipits is one of the many reasons for the decline in the cuckoo. So many of the really common birds of my country boyhood are in crisis. In England, tree sparrows have declined by 71 per cent, lapwings by 80 per cent, and those huge murmurations of starlings, which I used to watch heading north to the night warmth of Birmingham, are a thing of the past.

April and May are the months to listen to the dawn chorus, when male birds sing to attract females and mark out territory. By and large, the bigger and more tuneful the song the more likely the male bird is to attract a mate.

The concert begins at around 4.15, before dawn breaks over Merlin's Hill. To stand alone in a field in England and listen to the morning chorus of the birds is to remember why life is precious. I am in my dressing gown and wellingtons, unshaven, though none of the performers seem to care that I am inappropriately dressed, casual but unsmart. The birds sing in this order: the song thrush goes to the top of the ash and sings, to borrow Browning's words:

each song twice over
Lest you should think he never could recapture
The first fine careless rapture!

The song thrush is followed by a robin and a blackbird, also on the riverside, then the brown-barred wren by the newt ditch, the blue tits, the chaffinch, a dunnock, the blackcap, a pheasant, all against the persiflage of jackdaws who are cavorting in the sky above the derelict barn at the Grove. A skylark takes to the air, and two male meadow pipits also make singing ascensions.

I will proselytize on behalf of the dawn chorus. If you rise at dawn in May you can savour the world before the pandemonium din of the Industrial Revolution and 24/7 shopping.

There is now an International Dawn Chorus Day, which was founded courtesy of the Urban Wildlife Trust in Birmingham. This is international in the way the American football World Series is global. It's a British thing. As the journalist Henry Porter once pointed out, 'Whatever our self-denigration and decline, you cannot take away from the British a genius for the appreciation of nature, particularly birds.' We do seem to have been especially well appointed with birdie authors: Hudson, BB (Denys Watkins-Pitchford), Peter Scott, Viscount Grey of Fallodon, and J. A. Baker, the author of one drop-jaw classic, *The Peregrine*. Of course, some science Puritan will aver that British nature writing is diseased by 'species shift', or what W. H. Hudson (a leading practitioner) termed 'extra-natural' experience – the placing of the author inside the head and body of the being described. The same lab-coated lobby invariably sign off with the dig that 'nature writing' and, by extension, 'nature reading' are the habit of metropolitans detached from the real Nature of the red teeth and claws.

Every time I hear this argument I wind back my memory more than thirty years, to the little second sitting room of my grandparents' house in Withington. They had impeccable country credentials stretching back centuries, although admittedly in my grandfather's case only to the early 1600s. There were no parish records before then.

In the second sitting room, there are only three shelves of the dark wood bookcase; on them are a few respectable novels in paper polka-dot jackets (led by Du Maurier and Somerset Maugham), at least ten books about Herefordshire (I must have read *Where Wye and Severn Flow* twelve times by the age of twelve) . . . and an awful lot of books by Romany, aka the Reverend George Bramwell Evens, a BBC radio broadcaster and writer on nature. There was *Out with Romany*, *Out with Romany Again*, *Out with Romany by Meadow and Stream*, *Out with Romany Once More*, *Out with Romany by the Sea* . . .

There was nothing unusual about that little library. Everybody in the country had books on nature, farming and shooting, Brian Vesey-Fitzgerald for knowledge, James Herriot for laughs. And the worst anthropomorphizers of all are country people. I have never known a sow badger to be anything but an 'old girl', and when the gender of an animal is unknown it is always 'he', and never 'it'.

And I wonder, is it really so difficult to enter, in some slight degree, into the mind-frame of an animal? Are we not all beasts?

There's an evening chorus too, and it is best enjoyed on a day like this, when the light is seductive in white veils, and there is enough moisture in the dusk air to intensify the floral incense of the spring meadow. Two male blackbirds, on opposite sides of the field, one in the Grove hedge, one in Bank hedge, sing against each other in an ecstatic proclamation of their stake in the world.

Oh, the joy to be alive in England, in Meadowland, once May is here.

If merry May is the month for listening to the dawn chorus, it is also the time for fox-watching because the adults, with hungry cubs, are forced out in daylight, and the cubs themselves are up above ground playing. They are wholly incautious this evening, having slunk under the fence from the copse to rough-and-tumble in the mattress grass of the field. Their turquoise eyes watch me approach until I can be no more than thirty feet from them; only then do they scamper back to their earth.

Such unwariness will not last. In a month they will be nervous of me, a human, and they will have an awakened atavistic liking of the night. There are three of them, weaned and about eight weeks old.

I'm aware that the vixen is watching me watching them. She has emerged from the thicket with a mallard duckling dangling out of the corner of her mouth. A spiv with a fag would look less shifty.

Mallard ducklings are mainly brown with pale faces. Of the eight hatchlings born to the female who sat under the Elephant Tree just upstream, one was a garish Tweety yellow, which was the same as a death sentence.

The duckling is for the cubs. Their mother has been scoffing voles or rats, dug out from the river bank at the bottom of the thicket.

We are old acquaintances, the vixen and I, and she recognizes my face or maybe my smell. Anybody else and she would have warned the cubs minutes ago. As, indeed, she would have done if the dogs had been with me.

The cubs should make the most of their duck. Such are the difficulties of cubs' lives that by August insects and worms will be staple items of their diet. Carabidae (beetles), Lepidoptera (butterflies and moths), grasshoppers and crickets, slugs and snails, arachnids (spiders) and maggots will also be taken. And the lowlier the status of the fox, the more low-down invertebrates it will eat.

*

By 7 May the hawthorn hedge on the track at the top of the farm is in full creation-green leaf. Three days need to pass before the hedges around the meadow, at the bottom of the valley where the frost likes to live, are wholly green.

Only once do I venture to look at the curlew's nest, on this slow, close afternoon when she flies to stretch her wings. I have stared at the spot for hours, and know its location down to a yard, and still it takes me a minute to actually spot the eggs. But there they are, four of them, pear-shaped, a gorgeous avocado green blotched with brown. It's been three weeks since they were laid; they are only a week off hatching.

Little patches of foam are glued to the taller grass stalks. Cuckoo spit. On gently smearing out the foam on my outstretched fingers, I uncover the pale greeny-yellow naked being that lives inside – the nymph of the common frog-hopper, *Philaenus spumarius*. The so-called 'spit' is produced by the larva blowing bubbles from its anus, and serves to keep the creature moist and hidden from predators. After all, stripped from its frenzied foam the frog-hopper looks a tender treat to a carnivore. The frog-hopper is one of the true wonders of meadowland: the adult *Philaenus spumarius* is, millimetre for millimetre, the world's greatest jumper, leaping as high as 70cm – the equivalent of a human jumping over the Great Pyramid of Giza. To do this the bug attains an initial acceleration of some 4,000 metres per second.

John Clare was convinced that frog-hoppers had more attributes still:

> They begin in little white nottles of spittle on the backs of leaves and flowers. How they come I don't know but they are always seen plentiful in moist weather and are one of the shepherd's weather glasses. When the head of the insect is seen upward it is said to token fine weather; when downward, on the contrary wet may be expected.

Right in the bottom of the sward, in the tangly bases of perennial grasses and the accumulation of vegetative debris, there is another mighty jumper. This is the springtail, a tiny terrestrial jumping shrimp. Parting the grass to the red earth, I find one and touch it. The springtail does what it says on the label: it vaults, using a hydraulic piston on its underside which it drives into the ground for lift-off. There are about 250 *Collembola* species in Britain, and they represent an ancient group of primitive insects that bounded on the Earth 400 million years ago. Like the red soil, they are Devonian.

Down here, where my fingers are exploring hidden

micro-universes, and the soil is always moist to touch, the invertebrates exist in daunting numbers. Each acre of the meadow contains several hundred million insects. Together they weigh 0.2 tons. Or thereabouts.

10 MAY The doily leaves of buttercups are ever more evident. The blossom of the wild apple tree in the Grove hedge is pretty in pink. I can almost ignore the brutal rain. After days of mellow living, when I hubristically settled into the certainty of spring turning to summer in linear progression, May does its trick of turning down the thermostat. There is a dead lamb in a neighbour's field; the ravens and their one child are red-capped with blood from their gorging.

12 MAY The seeds in the grasses are thickening; the grass and flowers in the field are a crop, they have purpose, so I confine myself to walking the edges to avoid trampling. There is dew, luscious and anointing on the grass this morning. I am standing under the oak tree in Marsh Field hedge, in its cast tapestry of light-and-dark, when I see the caramel stoat, sitting up, looking at the hedge. He is in another time frame, playing the ancient game of killer. He twines into the hedge, and

twines out with a fledgling blackbird in his mouth. He never sees me.

W

14 MAY Another uproarious morning of birdsong to greet the rise of the light. The dawn chorus is also an aid to determining who is nesting and where. Male birds proclaim their ownership by singing from a conspicuous vantage point. In the field's hedges this morning there are three pairs of blue tits, two robins, two wrens, one song thrush, one long-tailed tit, two blackbirds, one great tit, one chaffinch, one hedge sparrow.

No fewer than thirty-four species of British birds commonly nest in hedges, the most typical of which is the hedge sparrow. Its name in Old English was *hege-sugge*. Today, it is called the dunnock, but it remains the little bird of the hedges. A hedge sparrow's sky-blue eggs huddled in a nest is one of the prettiest sights of spring; the picture the broken eggs make in the grass below Bank hedge is ugly. All the contents have been sucked, pecked or licked clean away. Two magpies have taken to loitering in the field. They nest in the solitary oak upstream of the quarry. One of them had its beak dipping in the egg as I walked in on the scene.

W

15 MAY The first red clover flowers appear, on which several species of bee are feeding, late into an evening of Tuscan light, with the desperation of *Titanic* survivors clinging to life rafts. The sorrel heads are already turning into rusty-red towers.

Sorrel, an upright perennial member of the dock family, likes pastureland untreated by chemicals. The plant's Latin name says it all: *rumex* was a type of Roman javelin; *acetosa* means roughly 'vinegar'. In other words, sour spear-shaped leaf. And tart it is. A kind of masochistic pleasure comes in chewing it. When haying, agricultural workers of old would bite on the leaves to stimulate saliva in their mouths. Used in medieval cooking the way we use lemon and lime, the plant's high oxalic content gives it its characteristic sharpness. Until the time of Henry VIII it was cultivated as a herb and used in 'green sauce' for fish. And now the spires of its flowers, which reach 60cm in height, impart a red mist to the field.

The red seeds are a food source for finches (goldfinches in particular), and the leaves make a meal for the caterpillars of the small copper butterfly.

At night, under gleaming stars, I stand at the edge of the meadow and inhale. I can smell the grass getting sweeter.

Mid-May, and the curlews have stopped their singing. I miss it so.

The curlews in the meadow are wise to keep their quiet; they have two chicks, black yarmulkes on their heads and some fancy black stripes across their eyes. They are being fed by both parents who, with meadowlife smarts, land about twenty yards from the nest and creep inconspicuously in on foot, only their crouched heads visible above the screen of rippling grass. Soon only the male will feed the gaping-mouthed young; the female curlew has done her bit. Curlews only have one brood per year.

On occasion, I realize, the curlew adults are not flying off for food but walking to the nearby newt ditch. From under the hedge I cannot see past the thistles to the ditch; on my next visit to the field I hide, standing still, in the shadows and the stands of hazel of the copse. A human tree. The wait is worth it for the joy of seeing the curlews' private dining: the curlews almost upend in the ditch, pulling out worms and taking insects off the surface, presumably skaters. More than once I see squiggling frogs and newts in their forceps of bills.

While I am watching the curlews feed their young I note that the foolish meadow pipit is not entirely without guile. The female is carrying the faecal sacs of the recently hatched young and dropping them under the hedge, so the smell of the excrement will not attract predators to the nest.

*

In a lazy-aired evening I dig up pignuts, whose feathery white heads gawk above the grass. The pignut is a member of the carrot family, and its tuber – which is round and cobnut-sized – is sweetly edible underneath its black husk. The knack in harvesting pignuts is to trace the thin stem down to ground level, then follow the immensely fragile long single root down to the tuber itself. Break the root thread on your 10–15cm journey down into the earth and you lose the tuber treasure. Caliban in *The Tempest* dug up the 'fairy potato' with his bare fingers; the red Devonian clay of Lower Meadow requires a spade to get through it.

By the time I have twenty pignuts in my carrier bag, it is getting dark and the swifts are screeching around the roof of the house, and the tawny owlets are wheezily demanding food in the old quarry. I'm about halfway back up the bank when there is an almighty whirring of wings in the grass. A red grouse rockets off. Only the bird brain of the grouse knows what it was doing here, a mile or more from its mountain-top home.

Another evening: I sit under the twin oaks, the sunlight creating Japanese willow pattern shading on the bank. I'm smelling the old brown brook, which is glugging in a noise curiously akin to water going down the plughole of the bath. Edith is swimming, head out of the water, as matrons of a certain age do. And mayflies are dropping by the dozen on to the water surface around her, spinning crazy circles, spinning themselves to death. Downstream in Periscope Pool I can hear the trout jumping. Edith emerges with sealskin shine, shakes herself and lies down beside me. It is warm, and the comfort of dogs is always reassuring.

I'm jerked to attention from my dozing by squeaking. Daubenton's bats are chasing down the mayflies, picking them off the water with their hobbit-hairy claws. On the brook's edge some red-faced, out-late swallows are collecting mud for their nests.

16 MAY Early murk, banished by an ascendant sun. Three trout lie like wooden clubs in Periscope Pool, faces upstream. They are the counterpoint to the frenzy of the rest of nature. From just after dawn, the chaffinches in Bank hedge have been feeding their four gape-mouthed hatchlings every two to three minutes. Green caterpillars are delivered in vice-beaks, borne by white-barred wings. So continuous is the activity that it becomes etched permanently on the side of the meadow scene.

When my Parry ancestors arrived in Herefordshire nine hundred years ago, and stood on the brow of the

Black Mountains and looked out over England, what did they see? A land not unlike now. There were already emerald meadows between the trees; the next village over, Maescoed, is *maes-y-coed*, meaning field in the wood, and was so named as early as 1139. The Wain farm along the lane draws its title from the Welsh for meadow, *gwaun*, and not the Middle English *wain*, meaning wagon.

The oldest hedge on the farm is eight hundred years old; carved from the wildwood in the Middle Ages, the fields have hardly changed their shape since. The Georgian enclosures did not affect the Welsh Marches as permanent pasture did not follow the common three-field system of fallow/winter corn/spring corn.

> *My wild field catalogue of flowers*
> *Grows in my rhymes as thick as showers*
> *Tedious and long as they may be*
> *To some, they never weary me.*
> John Clare, 'The Shepherd's Calendar'

The meadow buttercup, *Ranunculus acris*, is a serious inhabiter of pastureland and hay meadows, its abundance an index of the age of the grassland. Cattle usually avoid the plant because of its high ranunculin content, which inflames the digestive system if eaten raw, though it is trouble-free in hay. Beggars of yore used to blister their skin with buttercup juice to arouse the sympathy of passers-by, hence 'blister plant'; country people name the meadow buttercup the crowflower, because of its acridity. And because the crow is always the omen of evil. When he was eighty William Parry of Longtown told the Victorian folklorist Ella Leather about a shepherd who was attacked on the mountain by two brothers. The shepherd told the brothers, 'If you kill me, the very crows will cry out, and speak of it!' The brothers ignored the warning. Thereafter, they could not go out without being mobbed by crows. Their nerves stressed and stretched, they unmasked themselves by blurting out their sin. And were hanged.

The meadow buttercup flowers from May to August, and the first gold heads are shining loud, so that the low, crouching vixen appears to be wearing an elaborate Cleopatran crown. A small number of rabbits have hopped through from the Grove and are nibbling at the grass by the anthills. She has already had one go at the rabbits, rushing them, but the alarm was signalled by one thumping on the ground and they bolted to the burrows on the bank.

All she has done since is lie like a sphinx in the flowery mead, and wait for the rabbits to come back. When one wanders too close, she explodes to snatch it by the neck. She is a pretty killer.

The grass shimmies, then bows its head in racing waves before the wind. Someone has sprinkled caster sugar on the hedge. The 20th of May and the hawthorn has turned the world an eye-catching white. This is the white time. White for hawthorn blossom heaped on the hedges, white for the stitchwort growing under the hedge.

A fox has left a territorial scat on the stone floor of the Bank gateway. I can see earth in it. Last night and the night before it rained, and was warm, and hundreds of worms were crawling over the grass. I counted ten per metre. One of the foxes has made a meal of them, but the dirt in its stomach is indigestible, hence its appearance in the fox's excrement.

24 MAY No, the field is not always beautiful: the dandelion flowers have been turned, by the passage of time, into seedy, pale clocks. The white time: the field has all the allure of dandruff on a school blazer.

Note scrawled on paper, 25 May: 'While fixing some wire across gat in Marsh hedge I disturb a hedgehog suckling three young.' A gat is dialect for a gap, and stringing a piece of barbed wire across it is only marginally more industrious than getting the dog to sit in the hole and keep the cows from pushing through.

But the heat was beating, the clay gone to iron, so that fixing in fence posts did not appeal.

Herefordshire clay: it is either wet and sludgy, or hot and hard. There are about two days a year when you can work it sensibly. The heatwave has brought out the butterflies, and over the surface of the meadow there is now a constant interference of cabbage whites and meadow browns. I also see a blue butterfly I cannot identify, until I look it up in *The Observer Book of Butterflies*, given to me by my parents when I was nine. A female common blue.

On the cow parsley that sprawls into the meadow from the thicket, there are also orange-tip butterflies. White saucer blossoms of *Anthriscus sylvestris*, their wings closed vertically above their backs, the orange-tips are fantastically difficult to discern even though I am only inches away. The green-and-white mottling of their underwings is the acme of eye-fooling camouflage.

The sight of the adult orange-tips nectaring prompts me to check the cuckoo flowers by the ditch to see if their caterpillars are there. After some searching, I find five green orange-tip larvae. Cuckoo flower, along with garlic mustard, is the primary food source for orange-tip young, along with each other. The caterpillars are devout cannibals.

JUNE

Shrew

3 June All the trees are now fully dressed, including the ash.

Hovering above the luxuriant grass is the glow of gipsy-gold from the buttercups; my wellingtons are yellow from the flowers' pollen. The baby-blue air is breathless, only moved by the beat of the swallows' wings as they hawk midges over the field. But there is noise: the constant drone of hoverflies, the buzz of horseflies, the hum of bees.

The field looks different. Not just because I am sitting down in the wild triangle, with a bumblebee's view across a lake of grass and flowers, but when the meadow is full of flora it seems tighter and smaller, and is almost unrecognizable from the chill bleached space of winter. In the sunshine, meadow brown butterflies swarm over the grass, the males chasing off other males in their pursuit of a beguiling female.

In the shaded but desiccated land of the hedge bottom, where I am crouched, a dun shrew runs over my leg. She is careless of my presence and pokes around in the old leaves in an amphetamine frenzy. Over the next ten minutes this tiny, long-trunked mammal puts on a horror show, although one can only admire her murderous dexterity. She dismembers five beetles with rapid movements of her jaws, before rubbing and rolling a grey slug with her snout, presumably to tenderize it. Occasionally she nips it; her saliva contains a poison that immobilizes and eventually kills the victim. She also wolfs down woodlice, preferring the *Philoscia muscorum* louse to *Porcellio scaber*. Between courses she washes assiduously. No dunce, she refuses to snack on a large black beetle that looks capable of fighting back.

Eventually, she decides to head for home, somewhere out in the field. I follow her progress, parting the grass in her wake. Or, I should say, the flowers in her wake, because the midsummer sward is now a running floral riot of white stitchwort, gold dyer's greenweed, purple common vetch, blue bugle . . . I almost miss the shrew's minute burrow, which is next to a solitary oak seedling, intent on returning the field to forest.

She is a common shrew, *Sorex araneus*, at 6cm about 2cm bigger than the pygmy version. Shrews require gargantuan amounts of food due to a very high metabolic rate, and a shrew can eat its own body weight in twenty-four hours. So they are almost always hunting and eating, day and night, night and day. Mammalian predators rarely eat shrews because shrews have glands on their flanks which produce a foul odour. The Latin name *araneus* means spider; this refers to the old belief that shrews were poisonous, like spiders. Feathered raptors, however, make a principal meal of *Sorex araneus*. Most birds have no sense of smell.

Shrews mate from March, and up to four litters are produced a year. By sixteen days old the young begin to emerge from the nest, and are said to sometimes follow their mother around in a 'caravan', whereby a young shrew grabs the tail of the shrew in front of it, so the mother takes the lead and her offspring follow in a train.

I would like to see such a caravan. I never have.

A beautiful evening travels down the Bank Field and through the pignuts. Blue tits hop in and out of the hedge, cleavers slosh against its bottom. Wood pigeon are calling from the dead elm, 'Take-two-cows, taffy take-two'. The western sun bathes the land in gentle mythic pink. Even the bog-standard galvanized field gates glow enchantingly.

Almost the moment I reach the gate to the meadow a small jet engine starts up. Or so it sounds. The adult May bug, at 30mm long, is as easy to see as it is to hear. I duck. *Melolontha melolontha* has been slow to emerge this year of cold spring. So June bug then. Or cockchafer, spang beetle, or maybe billy witch, chovy, mitchamador, kittywitch and midsummer dor. Despite its fearsome size and needle-pointed rear, the cockchafer is harmless. To duck out of the way appears to be an involuntary, natural reflex. The cockchafer tanks past at head height minding its own businesses, which are sex and food.

Lying in the gateway like a chip off a varnished mahogany table, into which someone has etched stylish white triangles, is a downed cockchafer. Cockchafers are fatally attracted to light and glass, into which they hurtle at 11mph. This one probably hit the headlamps of the tractor last night. The cockchafer is not a bug but a beetle, and given its furry head and eccentric hand-like antennae a quite charming beetle. I pick it up and lay it in my open palm on its back. Its legs unfold like the Leatherman utility tool I always carry in my pocket. Perhaps it is dying of age rather than the trauma of a tractor crash. A cockchafer lives for a brief, brilliant six weeks.

There are other cockchafers gathering in the oaks, having just risen from their white-grub life under the grass, where they chew the roots in a clandestine four-year-long infancy, to then stagger into the sky on transparent wings. The grubs are outlandish, and curl into a distinctive crescent, thick and 4cm long, when uncovered. In some regions of Britain they are known as rookworms, because the rooks seek them above all other treats. Soon the female cockchafer will begin the cycle again, on some warm night like this, when she lays her eggs in the soil, using the pointy pygidium at the end of the abdomen, which is an implement for piercing the ground, not the human epidermis.

For an hour I sweat away in Bank Field restringing the fence along the river which the sheep are determined to push down with their rubbing to relieve itches. (The ovine way of asking to be sheared.) By the time I finish the fencing the noctule bats are seeking the lumbering cockchafers. Noctules are the peregrine falcons of the Chiroptera order. On narrow wings that measure nearly fourteen inches across, the noctules fly high over the meadow into the first stars. Then free-fall stoop. Noctules can eat on the wing. As they flutter up on their clockwork wings above me I can hear, I am sure, the sound of showering cockchafer scales.

The noctule is the largest bat in the country and one of the few prepared to fly in open spaces. (About 10 per cent of bats are eaten by birds of prey.) Other bats are beginning to flicker into the night. Under the river alders the Daubenton is at work. Against the last light behind the mountain I can make out bats hunting down Marsh Field hedge and in among the cows. These are greater horseshoe bats after dungflies.

June thunder. Swallows swoop in brief white whirls over a prematurely darkening meadow, always keeping low, their mouths nets to catch the congregation of insects forced down by the weather. Lightning jigs on the mountain. A whip cracks somewhere.

Then the rain comes, heavy raindrops crashing

through the tiles of oak. The fluorescent florets of the hogweed and cow parsley are beaten down; it is night at mid-evening. A fox – one of the young ones – emerges from the Grove ditch, and I think it is going to hunt rabbits but instead it rushes along the rain-lashed edge of the meadow to the earth and the dry. The fox cubs are roaming further and further afield. But on a day like this, home calls.

In the wreckage of the evening a heron lands and stabs at something at the Grove end of the field. I cannot see what it is, only that it is being eaten; only that it is large; a baby rabbit or a rat, something of that order. On its outsize wings the heron lifts into a sky still angry, to continue on its stately patrol. The newt ditch is swollen with rain; a common newt (*Triturus vulgaris*) cruises in slo-mo eating tadpoles; one outsize tadpole jams in the newt's mouth, and only after terrier-type head-shaking can the spotty aquatic lizard gulp its cousin down.

The ripe grass, heavy with seed, has been flayed flat by the violence of the wind and the rain. Somehow the grass, or most of it, lifts its head from its battering. The thin lance heads of the rye are most dismissive of wet; the fluffy heads of the cock's foot and vertical pagodas of crested dog's tail take longer to rise.

W

9 JUNE I am reading Viscount Grey's 1927 *The Charm of Birds*. Grey is the Foreign Secretary who took us into the Great War, the man who provided the epitaph for the prelapsarian continent: 'The lamps are going out all over Europe. We shall not see them lit again in our lifetime.' Grey was a reluctant politician, and was always happier birding than engaging with matters of state. On 9 June 1910, however, Grey managed to mix business with pleasure, taking Theodore Roosevelt, the ex-President of the USA, on a 'bird walk' down the Itchen valley. During their walk they saw forty separate species of bird.

I am troubled by the obvious question: how many would the walk provide today?

10 JUNE One of our cows has escaped into the meadow; she is not anxious to return to the leaner pickings of Marsh Field, and displays some tantrumy head-tossing when I start herding her back. There is a funny art to cow-herding: if you stand behind a cow and stick your left arm out it will go right, and vice versa. After some windmill signalling she steadies into a sensible line of passage. Cows are not stupid; she has been caught playing truant, and knows the game is up. She walks deflatedly towards the gate into Marsh Field, me behind her, the lonely wight in the eternal picture of

a man herding a cow. We are acting out time-honoured roles, and there is a kind of unspoken companionship in our journey. I pull up a grass stalk (timothy) to chew, to make the rustic simplicity complete.

A desperate squealing in the grass. A field mouse's nest, a ball of grass, absolutely dry despite the morning's rain, has been spliced open by the cow's hoof to reveal three blind, brown, sugar mice. One baby has been squashed open, and squeezed from its skin. I cover the nest as best I can, and flick away the bloody corpse with my wellington.

A cow exerts about 1.5kg per square centimetre. I have had every bone in my feet broken by careless cows. What hope a naked newborn mouse?

But a cow's foot is not all bad news. The hoofprint provides a microclimate that specialized invertebrae such as the blue adonis butterfly require for oviposition.

12 JUNE The blossom on the hawthorn in the hedges has decayed to a tentative cerise. White moths float around by day and by night. There are frogs in the grass, where it is deepest and retains moisture even at midday. Elderflowers are out on the east and south sides of the meadow hedge, as is the honeysuckle. Bird's-foot trefoil blooms red-and-yellow in the understorey of the sward, the bacon and egg plant. Clouds

of pollen hang above the meadow. When it is cool in the evening, brown and black slugs slide along the grass. The purple-headed thistles in the entrances where the cows like to stand and stare are a metre high.

Keith Probert comes down the track wanting to know if I'd like to borrow his new Hereford bull. 'Bought him last week. The lads been giving me stick about him.' No one can believe that a cattle farmer has bought a Hereford. A Belgian Blue or a Simmental would be a much better commercial proposition.

I can understand why. Belgian Blues are over-developed musclemen and Simmentals have the personality of a machine. A Hereford is tradition, is companionship, is a bit of the old ways.

How lovely it is to lie in a field and dream. I lie on my back in a casual crucifix, which seems an instinctive shape, since it is both arms-wide welcome and submission before Nature. To lie with arms straight by one's side is the posture of death, the attitude of the coffin.

Above me the skylark flutters into the haze, all the while singing a silken tent over its territory, until it is a speck in my eye.

The lark is a male defending his patch. But where is his mate's nest?

It takes earthbound me an hour to find the lark's house, tunnelled into a clump of grass. With exquisite nervous care I pull back blades to uncover three brown speckled eggs. A sort of treasure.

It starts young, the obsession with Nature. As a ten-year-old I remember roaming for hours with my cousins and a dog, climbing trees to reach the pure porcelain eggs of wood pigeons. The first school photograph of me shows a Young Ornithologists' Club (YOC) blue cloth badge visible on my jumper, revealed by a seditious pull back of the blazer lapel. The first thing I ever had published was in the YOC magazine *Bird Life*. (The next was in the *Guardian*, but hey ho.) My long-suffering parents took me time and time again to Slimbridge Wildfowl Trust. My bedroom was a museum of preserved skeletons, beaks and feet, my proudest display being the beak of a puffin, found on the beach at Borth. In a clear plastic case, formerly used to pack the Timex watch that was obligatory for a boy in the 1970s, it looked like a particularly exotic brooch.

The meadow brown, *Maniola jurtina*, is on the wing despite the lichen-dull weather. A medium-sized butterfly, the meadow brown is distinguished by orange patches containing one 'eyespot' on the forewings. The caterpillars feed on a variety of native grasses such as bents, fescues, cock's foot and meadow grasses. They do not wander far afield if they can help it and the meadow browns mating and flying today may never leave these five acres.

Two meadow browns are mating face to face, missionary style, head uppermost, on a nettle leaf. They make a perfect heart shape.

W

The fox cubs are cautious now, neophobic, averse to daylight, wary of me. For a week or more I do not spot them. Then Edith flushes one of them out from the thicket, which is one of the usual fox places for lying up. Foxes spend very little time below ground, outside of birth and a hard winter. The slimness of youth allows the cub to speed under the fence into the field; the matronly Edith needs to find a larger hole, by which time the fox is nothing but a fading impression on the retina.

A sharper image comes to mind, of Sniffy, our deceased miniature Jack Russell, who chased the dog fox out from the same spot on a declining dreary October afternoon years back. The little dog gave full chase to the big dog. Neither of them stopped to ponder the ridiculousness of it all. The fox was the bigger by a factor of ten.

The fox family is not alone in lying above ground. Some of the young rabbits have made forms in the long grass by the Grove bank. They graze watchfully in the afternoon, before playing the chase games beloved of many mammalian young.

19 JUNE It is almost midsummer, and the light makes one resist sleep. There is daylight seventeen hours a day, double that of midwinter. Eternity would not be long enough if it was composed of English summer eves like this. I decide to take a turn around the fields. Walking down Marsh Field I can make out the shape of two animals standing together on the far side of the grass-sea meadow. A fox and a badger have met on their shared path; the badger is motionless, the fox is jutting its head forward, its gravel-gargle of anger audible thirty yards away. The badger is unimpressed, and it is the fox that gives way, bounding off into the meadow and around the black-and-white Buddha.

20 JUNE There is a dead velvety shrew near my lying-up place in the wild triangle. The body is still warm. On the neck I can just discern tiny bite marks in the wet-tened fur. Shrews are territorial to the death.

*

Yellow rattle really does rattle. While some regional names for the plant point to the similarity the shaken seeds in their pods have to the sounds of child's play (baby's rattle) or money tossed about in a bag (shepherd's purse), Herefordians have a bleaker aural association for *Rhinanthus minor*. Locally it is rochlis, the death rattle. And it does bring death of a kind. Strictly speaking, *Rhinanthus minor* is a hemi-parasite, and, while it can photosynthesize, it is happiest when its roots grasp those of grasses so it can suck the life out of them. In the creation of a wildflower meadow, yellow rattle is as close to indispensable as it gets. As well as controlling the vigour of grasses, yellow rattle produces springtime yellow flowers which are attractive to bees. It only grows intermittently in the field, in the bare dry patch in the middle, and the northern end.

When I accidentally disturb the meadow pipits' nest the female flies up and gives me the full works of distraction, fluttering along the top of the grass for five yards or so to a thin patch, where she lies prostrate with one wing held out 'injured' and her tail fanned to the ground. I don't wish to disappoint her, and follow. At which point she flies over the hedge into Marsh Field with a tinkling laugh.

*

Some years ago, I betrayed my own rule, which is you have to like your livestock. So I bought, to add to the Ryelands and Shetlands, twelve Hebridean sheep – small, black, horned, primitive. And cloned. Of the black dozen I have only ever managed to like Hilda, the lead sheep, she of the retroussé Michael Jackson nose and unsated stomach. A neighbour donated a ram, called . . . Rammy. The Hebridean is a productive little sheep; on halfway decent and varied grazing like Lower Meadow it breeds well, is healthy, is flavoursome in the flesh, and produces a lustrous coat which fetches a good price with the Wool Board and private purchasers alike. The breed also scores highly with landowners intent on environmentally friendly, subsidy-attractive 'conservation grazing'.

But they escape, and jump like deer. I have them in the shed for shearing and one leaps the holding pen into the pen where I am wielding the electric shears. She then tries to vault over me, and only by diving to the side do I avoid her head, which is made from tombstone.

Shearing is fine in your twenties; in your forties it kills. The definition of 'back-breaking' should be 'a prolonged period of shearing sheep in the New Zealand position'.

Almost everyone shears the New Zealand way, which is to put the sheep on its arse, back to one's legs, and shear down with electric clippers.

I start off at a reasonable(ish) rate of a sheep every two minutes, the clippers neatly sliding under the line of yellow risen lanolin in the fleece; by sheep number 22 I am down to a sheep every five minutes, and starting to make 'double blows', or two shears, because the first isn't close enough. I also nick one ewe's skin badly, and have to blast her with purple spray, or aerosolized gentian violet. By sheep number 26 I am 140 years old. By sheep number 31 I cheat. I park the tractor on the track so no one can surprise me, and shear the rest of the flock with the sheep standing up, a halter around their heads, tied to a gate. I sit down to do it.

But I can never tell anyone about it because it is so seriously uncool.

My back is broken, and the exertion turned me into the portrait in Dorian Gray's attic. My hands, though, are baby soft from the lanolin in the fleeces.

On Midsummer's Eve one of the strangest moments of my life befalls me. At ten, when I go out to shut the chickens in their hut in the paddock, the demi-light is alive with magick. But then Midsummer is always a night for Puck, fairies and wonder; I once heard a nightingale sing in the valley at Midsummer, the only time I have.

No sooner have I closed the pop-hole than the

three horses and donkey appear noiselessly out of the hedge-shadows, and start circling me in a carousel of prancing and rearing. They go round and round me, faster and faster. Wilder. Wilder. Frankly, I am frightened; the wind from George's lashing hooves crosses my face. The trotting donkey is the slowest in the circle, the comma mark, and I dash in front of her and out of the paddock, over the field gate in a jumping style I last used on a school sports day decades ago. The Fosbury Flop.

The equines continue circling the hut, until Zeb, my horse, breaks out of the circle and runs, with absolute deliberation, up to me. And gently tugs on the sleeve of my shirt. He pulls the sleeve again, with infinite courtesy.

Animals can, of course, talk. In that moment he and I are one, indivisible. I can see inside his great impenetrable chestnut head, see every slow bestial process. I am a fellow animal and he wants me to play.

I kiss his head in apology. And he is away back to the merry-go-round.

The enchantments have not finished, for Snowdrop the donkey also trots up and pulls my sleeve with her lips, wanting me to play. She too gets a kiss, and then waddles back to join the fairground show in the lumined aethereality.

There is a new noise in the field. The meadow grasshopper (*Chorthippus parallelus*) is coming into song, only here and there, a brief interference. The song of this little green locust is produced by stridulation, the rubbing of a file on the inside of the hind legs against the forewings. The grasshoppers 'chirp' – so much nicer a descriptor – to each other in the sun. The male is louder and more persistent than the female. The grasshopper is more than a musician; it is an important protein for meadowland predators.

A crow is hopping about with evil intent. But the bird's desire is not matched by its physique. The grasshoppers bend their knees so cuticles on the hind legs slip into springs; when the muscle is relaxed, the energy catapults them into a new part of the grass forest. All the crow can do is watch as grasshoppers spring past. Grasshoppers have lineage stretching back 300 million years to the Carboniferous period; they are another landowner with a prior claim to humans.

27 JUNE Under the two old shading apple trees of Bank Field the cows are standing waiting for Constable to paint them.

I am on my stock-checking round on a June day

of Spanish heat, and pause to catch my breath by the river, which is running mountain-clean and clear over the green and pink pebbles. There is an attractive spread of golden-leaved saxifrage along the bank. On the red bank behind the water, kingfishers have dug a Chaco canyon of holes over the years; one orifice, the home in occupation, is spilling putrid black slime. The kingfisher's notion of removing excrement from the nest is a slovenly push out of the front door.

While I am in my trance, around the bend of the river comes a brown torpedo firing through the water, scanning as it comes. A lithe twist, an agile spin, a gymnastic roll, then the otter clambers out of the water on to the shingle directly in front of me.

There is something about this twenty-yard stretch of river, with its red cliffette behind and its overlapping alder trees, that encourages us all to consider it our own private space. The otter is no exception. It nuzzles at its chest and performs its ablutions.

On the 'how' of watching animals the nineteenth-century nature writer Richard Jefferies was correct when he stated in *The Gamekeeper at Home*:

> This is the secret of observation; stillness, silence, and apparent indifference. In some instinctive way these wild creatures learn to distinguish when one is or is not intent upon them in a

> spirit of enmity; and if very near, it is always the eye they watch. So long as you observe them as it were, from the corner of the eyeball, sideways, or look over their heads at something beyond, it is well.

As a good student of Jefferies, I stare above the otter's head; I am so close I can see each of its whiskers, each dripping bauble of water.

Unfortunately, the Jefferies rule of observation of wild creatures is unknown to Rupert the Border terrier, tied to my hand by pink baler twine. I can feel him tensing. In all likelihood he is staring. And baring his teeth.

After a minute or two of toilet the otter suddenly stops and looks around. And sees us. With some considerable self-possession it ambles across the stones in the shallows, and clambers up the far bank. The otter has not quite run away; in the military euphemism it has retreated. In water the otter is menacing, powerful, intentional. On land it is a lolloping toffee-brown dachshund.

I am strangely put in mind of dapper Edwardian gentlemen. (You can anthropomorphize otters as easily as moles.) And then, disturbingly, of Bristol Zoo, which is the only place where I have ever seen an otter so close. In a moment of unpleasant realization I understand that viewing the otter in the tank at Bristol

Zoo diminished this sighting in the wild. I saw the copy before the real thing. I saw the manufactured spectacle before the natural sighting.

Is this not what happens to us all today? Has *Autumnwatch* not killed the experience of being an amateur naturalist?

Or perhaps I just read Walter Benjamin's 'Work of Art in the Age of Mechanical Reproduction' too often when I was a History postgraduate?

A black wolf spider creeps from a crack in the ground of the hedge bottom. And starts off into the grass with deliberation. A male *Pardosa amentata*, I think. Wolf spiders do not spin webs; they hunt. On its head there are eight eyes arranged in three rows, all the better to see with; the first row comprises four small eyes, the second contains two larger eyes and the third row has two medium-sized eyes. (The four small anterior eyes are difficult to spot without a magnifying glass.) His peepers are better than mine. He sees her before I do. He stops and taps the ground with his front feet; she stills. Over the next minute, he cautiously scuttles, stops, scuttles until he is facing her. Then he begins waving the palps to the side of his face as though he was a sailor particularly weak at semaphoring. So begins the courtship in the fetid, closed-in world of the sward bottom.

His wariness is wise. Among spiders, the female of the species really is more deadly than the male. Lady arachnids have a well-deserved reputation for polishing off their suitors, *post copula*. But it has never been clear why this happens. Some biologists believe it is simply a mixture of female hunger and the availability of a meal that is in no position to run away. Others suspect that the male is actually sacrificing his life for the good of his genes. In other words, in becoming a meal for his paramour he somehow helps the offspring of their union.

He taps with his feet. Although she stares malevolently, the vibrations he sends must be good. This shadow foreplay can take hours. Sometimes, like today, it takes minutes. She accepts him. He rushes off. Wisely. And, ridiculously, I breathe a sympathetic sigh of relief for him.

Others have already mated. There are females basking in sun spots, where the grass has been parted by wind or animal activity, warming up their eggs which are attached to a hideous pod on their rear. A crusted molehill is much sought after for this pre-natal sunbathing.

I decide to go on a spider hunt, armed with a magnifying glass Sherlock Holmes would admire. And find: in the newt ditch the walk-on-water wolf spider, *Pirata piraticus*; in the grass and hedge there is also *Pisaura mirabilis*, a comparatively large wolf spider, the male of which woos his mate by offering her a dead fly or other insect, wrapped up in a silk parcel. *Pachygnatha degeeri* (large-jawed spider); *Clubiona reclusa* (silk cell spider); *Clubiona lutescens* (silk cell spider); *Lepthyphantes ericaeus* (money spider); *Lepthyphantes tenuis* (money spider); *Dismodicus bifrons* (money spider), which is a particularly common denizen of grasslands, found between the upper strata of field layer vegetation and litter at ground level; *Gongylidiellum vivum* (money spider); *Meioneta rurestris* (money spider). The money spider, in its various sub-species, is one of the commonest spiders living in grasslands. By my estimate I am sharing the meadow with over two million of these incy wincy spiders, few of which are more than 5mm long. They in turn are consuming over 230kg of invertebrates. For all the industry of the money spider in the meadow habitat, it delivers an elegant murder, tying up its victim in silken threads before dispatching it with a poisoned bite.

The spiders travel on the self-same silk threads, a personal flying carpet.

28 JUNE Under a chattering swallow-sky I run down the bank. Two of the Gloucester Old Spots have done a bunk from the orchard. Like the truant cow they

have headed for the luxury grass of Lower Meadow, where they have snouted the entrance gate off its hinges, and are now energetically eating, their mouths an epileptic, frothy green. They are pigs in clover.

When Freda was younger, I think about eight, we mislaid her. Forty acres is at its vastest when you cannot find your child, and there is a river all along the eastern boundary and a lane all along the western one. A lane with cars.

Freda went missing just before noon, on a day when the sun seemed locked above our heads and the land was holding its breath. Penny – who is less prone to panic – started a methodical search of the house and garden, while I speed-walked down the fields to the river. Soon I was running. And calling. At every alien shape in the water – a snagged plastic feed bag, a broken galvanized bucket washed down from God knows where – I imagined the worst.

Nothing. Sheening with sweat, I began running uphill in wellingtons – a feat usually beyond me – and decided to cut through the pig paddock to get to the fields leading to the lane. As I scrambled down the metal gate into the bare earth of the pigs' enclosure I saw, out of the corner of my eye, Freda's clothes entangled with a row of pink pigs, lying like sausages in the packet.

I can tell you what the end of the world looks like. In a circle around you everything dissolves and melts so that you know that life is an illusion, a pretty screen over the eternal expanding chaos of the universe. For one terrifying second I thought that the pigs had eaten Freda.

As I stumbled forward I could see that Freda was still inside the clothes. I could see that she was intact. As I reached her and touched her beautiful, rosy-cheeked face I could see that she was breathing. Colour burst into the world. Time sped up to its proper dimension. I'm probably imagining this, but I'm sure birds started singing too. Freda was fast asleep between two pigs. Feeling my fingers on her face, she opened her eyes. 'Hi, Daddy,' she said, before turning over on her side, the better to cuddle the pig next to her. The pig grunted its minor annoyance, before shifting its weight to accommodate her, setting off a porcine ripple as every pig adjusted its place in the sun.

I have another memory of pigs. A memory from my own childhood. I'm about six, standing on a wooden Davies Brooks lemonade box, arms leaning on the concrete wall of the pigsty at my grandparents' house. The pigs are milling around, squealing in excitement because they can smell the buckets of warm mash, boiled up from kitchen leftovers, that Poppop, as we call my grandfather, is about to tip into their metal troughs. As the food arcs down from the buckets I slyly look at my grandfather's thin arms, leather-brown below his rolled-up shirt sleeves; his arms always fascinate me, because the tendons – after fifty years of farming – are taut steel-wire hawsers.

The pigs jostle and barge, so that the herd order – for pigs are strictly hierarchical – is maintained and the top pigs have what they presume to be the biggest and best portions. 'Always remember this about pigs, Jahn,' he says, and suddenly jabs one pig on the ear with a spade. There is a grunt from the recesses of time, from the primordial swamp, as the pig bites at the shovel. Poppop retracts the shovel, bends down and points to the blade, swivelling it slightly so it catches the morning light. The pig has left gashing teeth marks in the steel. My grandfather is a man of few words but actions speak louder. Any animal that can leave bite marks in steel can bite off a human limb.

The problem with pigs is that one is never quite sure how they will react, dozily pacific or violently aggressive. The bristly Gloucester Old Spots do not like being shoved off the grass, and one whips around and tries to bite me. A shark's mouth is tender, quaint in comparison.

By the time I have got them back into their paddock, they have been out in the Aztec sun for too long and their pale ears are reddening with sunburn. They snort satisfiedly when I rub sun lotion into their ears.

The Gloucester Old Spots and the truant cow are not alone in their liking for my grass. In the greyscale evening a snuffling family party of badgers dines on the stuff too.

W

Young jackdaws fledged from the ruined barn at the Grove skirr in the sky. A kestrel fans the meadow; sunlight strikes through the goat willow, making the predator invisible. Below are field voles running through tunnels in the grass, dribbling urine as they go. The hawk, it is thought, can see the ultraviolet light reflected off the urine. The trees send shadows downhill on to the sighing grass. Hoverflies zip to the sudsy heads of cow parsley. Cabbage whites flock on the thistle heads. Grasshoppers and bees chant to the breeze, and the birds chorus back. The entire landscape is in motion.

Except for the grass snake – lying peacefully on a rock slab by the gate, and looking strangely unreal. This is the only time I see the snake in this entire year. When I look back to see if the hawk is still hovering I cannot make it out against the shadow play of the sun in the willows. Then when I look again to the snake it is no more.

W

29 JUNE Clouds race their shadows down the

mountain, across the valley, across the meadow, and up the criss-cross of fields on Merlin's Hill. Something else is stirring, moving. There are so many little shiny frogs in the newt ditch that they make the sedge grass in and around it tremulous.

30 JUNE The time for mowing is nigh. The hay-cut is when farmers wheel out their vintage tractors from dusty barn corners, and I am no exception. I spend the day servicing our 1978 International 474. And the finger bar mower, which looks for all the world like a horizontal garden hedge-cutter, which attaches to the tractor's rear. And the baler too. A long day of grease and spanners. The baby blue tits which live in the crack in the breeze blocks of the stables are companionable, though. They have not yet learned fear, and cheer me on.

JULY

Devil's bit scabious

ALL BIRDSONG HAS stopped; the noise of the field is a low insectoid drone.

I take a survey of the grasses: common bent; perennial rye grass; crested dog's tail; meadow foxtail; sweet vernal; quaking grass; red fescue; cock's foot; timothy; rough meadow grass. Where there has historically been grazing only: tufted hair-grass. A good deal of the grass has gone past seed, to leave empty purses. The hay would be better as fodder if it was headed, but cutting when the wildflowers have set seed means those wildflowers have a chance to reproduce because their seed is cast as a by-action of the mowing process.

2 JULY The morning mist is shattered by five green woodpeckers, a family party, exploding out of the promontory. One of the flying jewels laughs madly as it goes. Green woodpeckers feed upon worms and ground insects; the sparse light-deprived grass under the alders is pitted by the marks of their bills.

Like everyone who works the land I see auguries in

living things. The green woodpecker is the 'rain bird' of British folklore; in France it is still called *pic de la pluie* and its mocking cry known to herald the storm:

Lorsque le pivert crie
Il annonce la pluie.

According to the ornithologist Edward Allworthy Armstrong the green woodpecker was once the subject of a Neolithic cult, with woodpecker worship being superseded by other religions and eventually Christianity. Some trace element lingered on in the minds of men, for stories about the green woodpecker going against God's commandments are widespread. One German folk tale tells how the *Picus viridis* refused God's command to dig a well because it would spoil his gorgeous green-and-red plumage. As punishment, the bird was forbidden to ever drink from a pool or stream. Instead the green woodpecker must endlessly call for rain and fly into the air to receive the slaking drops.

Later: a thunderstorm. Doubtless whipped up by the green woodpeckers.

3 JULY The six-spot burnet moths are hatching. What urge persuaded the caterpillars back in spring to crawl up grass stems, spin papyrus cocoons around themselves, and assume there would be time in which to take wing? While I ponder the ineffability of it all, a

metamorphosed creature crawls from its Expressionist cabinet in the sward; it is a decrepit black being, and impossible to relate to the chubby yellow caterpillar that entered pupation. The afternoon sun makes the moth beautiful, its wings dry and expand, and the crimson spots that give this day-flying moth its name become visible. Except that the spots are not exactly crimson; the spots are more scarlet, the scarlet of *Cabaret* and Berlin brothels. The red lights on the moth's wings are not just a come-on to other burnets, they advertise the being's inedibility. Burnet moth caterpillars absorb hydrogen cyanide (HCN) from the glucosides in their principal food plant, bird's-foot trefoil. The HCN is retained during pupation into adulthood. The moth's gaudy dress warns that it is unpalatable, maybe downright deadly. (In the jargon, this warning coloration is 'aposematism'.) As the six-spot burnets emerge into this blissful afternoon of boundless hope, so do the flowers of their beloved thistles reach their purple peak; bird's-foot trefoil for their forthcoming young is already in bloom in the bottom of the sward. Everything is in perfect, synchronized order.

The meadow does a good line in thistles, though I try to restrict them to a five-foot-wide stand along the north end of Marsh Field hedge. Thistles have distinct Lebensraum inclinations. Dotted around the meadow, especially in the finger and by the newt ditch, is marsh thistle, a biennial which is hard to ignore: it grows to a metre and a half in height. A particularly splendid example is drenched in cabbage whites, which fuel themselves on nectar before floating off dreamily in search of a mate.

There are so many thistles in the gateway to Marsh Field I doubt I'll be able to open it. I have waited until now to chop them down with a hook, because the most ancient rule of British farming is this:

Cut thistles in May
They grow in a day
Cut them in June
That is too soon;
Cut them in July,
Then they will die.

9 JULY I take Edith with me to shoot wood pigeons on a glistening, curiously electric afternoon rising out of a sodden morning. The high wet grass uncomfortably soaks my jeans above my wellingtons. Patrolling the field edge under the sheltering twin oaks, Edith, bedraggled to her neck, suddenly stops, with her hackles springing cartoonishly upright. She's spotted one of the fox cubs – well a juvenile, now – fast asleep,

nose tucked into brush, on a parched ledge between the roots of the oaks. A fluffy carmine cushion.

To Edith's disappointment, we let sleeping foxes lie. The pigeons also turn out to be safe from harm; I don't get within fifty yards before they clapper out of the copse.

W

12 JULY High summer and one can hear the universe; so overwhelming is the accumulated sound of growing in the meadow and in hedges, of pollen being released, of particles moving in the heat, that all the minute motions together create a continuous hum: the sound of summer.

Meanwhile swifts tear the fabric of the sky on scything wings. The yarrow flowers are tall, the hawthorn flowers have turned to hard green haws. Blackfly fasten on the thistles, so too slender marmalade soldier beetles (*Rhagonycha fulva*) mating, tail to tail.

In the sitting room my son has left a pile of photos on the sofa; one shows him and his schoolfriends holding a daisy chain of prodigious length. Which makes my mind wander to flower culture.

Many field flowers used to be regarded by girls as love charms. Daisy petals were plucked to the rhyme 'He loves me, he loves me not'. Picking the grains from the rye grass was used for the rhyming verse designed to find the nature of your future husband: 'Tinker, tailor, soldier, sailor, rich man, poor man, beggar man, thief . . .' Field scabious (*Knautia arvensis*) buds were each given the name of a suitor, and the first to open was the man who would become your husband. Knapweed (*Centaurea nigra*) was stripped of its flowers, tucked between the breasts, and if in the morning it had regrown blossom your love was true.

Slightly more sophisticated, one feels, was *The Language of Flowers*, a guide to the means of communicating secret feelings through the sending of flowers. Published by Charlotte de la Tour in France in 1819, it was madly popular. Queen Victoria herself wore ivy leaves in her hair to symbolize fidelity to Albert.

There are girls' names, of course: Daisy, Poppy, Primrose.

There are flower games, such as do you like butter? Shining a buttercup under the chin to determine whether the subject liked butter. A tall buttercup flower against one's neck on the night of a full moon, or simply smelling the flower, causes insanity, hence the folk name 'crazy' or 'crazy bet'.

Cleavers were stuck to the back of blazers.

And my favourite: a thick blade of grass between pressed thumbs so it forms a reed, which is blown by the mouth. The noise, depending on delicacy, is either a raspberry, a curlewesque wail, or the tuba in Beethoven's Fifth.

The deep veins on the leaves of the plantain have earned the species the name 'ribwort'. Alternative names reflect the use of the stubby black flower head in a game akin to conkers, among them 'soldiers' and 'fighters'. Yet other names refer to the fact that farmers used to judge whether a haystack would be likely to catch fire by feeling a leaf of ribwort plantain to see how much moisture was left in the hay. Thus 'fire-leaf' and 'fire-weed'.

Pollen analysis has shown that ribwort plantain spread as Neolithic farming increased and the wild forest decreased. I cannot help but assume that Neolithic farmers found plantain as useful as I do in determining when to mow the grass for hay. When the plantain head is good enough to play soldiers the grass is good enough to cut. And after all, this is July, the month for which the medieval calendar advised 'With my scythe my meads I mow'.

16 July Under the hazels in the copse a fox (the vixen, I think) sits washing its front legs, a small red ember in the dying sun. Ten yards away a rabbit sits on top of an anthill, wholly in the view of the fox. The rabbit is also washing itself, paws to face. They ignore each

other. And the lion shall lie down with the lamb, the fox with the rabbit on this fantastic honeysuckled evening.

W

19 JULY Flying Ant Day. Out of the nests in Lower Meadow and Bank Field thousands and thousands of winged meadow ants are pushed into the balmy afternoon. This is an orchestrated Republican revolution; despite being two hundred yards apart, the prole ants in both sets of nests eject the winged queens and their winged male consorts at almost exactly the same time, 5.15pm.

The tops of the mounds seethe with insects, which groggily take to the air in a nuptial flight whereby males chase the queens for sky-high mating. The 100 Metre Club. Soon the winged ants are rising in smoky plumes, and are flying and landing everywhere, in my hair, on my arms, and the more I brush away, the more they seem to land. The plumes dissipate to leave the drifting air over the meadow tinged with grey.

The watchmaker's synchronized timing between the nests is a clever wheeze. With ants from several different colonies in the air at once, the incidence of inbreeding is lessened. Also, predators cannot cope with the sudden glut of food; pied wagtails are taking their haul of the insects, the spotted flycatcher is leaping off the fence at the stream of ants passing before her, and the sky is beginning to swirl with swifts, swallows and house martins. But the birds can only eat so many.

Some fertilized queens will survive to found a new colony somewhere. And nothing in nature is wasted. The bodies of the dead meadow ants will go to nourish the soil of the meadow. Ashes to ashes. Dust to dust. Flesh to flesh.

The yarrow heads in the grass are white, discs of intricate clustered jewellery. But it is the extraordinary leaves that are the key to identification, these being long, multi-divided, thus *millefolium* or 'thousand leaf' as part of its Latin name. There is probably more folklore attached to yarrow than any other plant. For the ancient Greeks it was the plant Achilles used to bind the wounds of his troops at Troy; for the Chinese the counting of dried yarrow stalks was held to be the means of divining in the I Ching, the Book of Changes. The Celts were convinced that the herb had psychotropic tendencies that allowed the imbiber to see their future spouse. The medieval English were altogether more down-to-earth: the bitterness of yarrow was found to be ideal in the flavouring of ale.

I pick it because it makes a rather refreshing tea.

W

Mowing grass for hay always brings a headache, especially if mowing an old-fashioned meadow the old-fashioned way. Not until the third week of July are the curlew fledglings grown enough to take wing; I see them leave one morning when I'm in Road Field and happen to look down on the meadow to see four curlews circle around, and then come up over my head, line astern, and out over the mountain, going steadily west. They are unconscious of me; this, however, does not stop me stupidly suffering the pride and sadness of a parent at their departure.

There are other wild things to take account of in the timing of the hay cut. Not until the last week is the yellow rattle seed properly set, and this is one of the plants I wish to encourage. Then there is a skylark still sitting but I peg out a plot around her so she will be left unharmed in an unmown island. The meadow pipit is on her second brood, so she too gets a private island.

At no other time is there a greater disparity between the superficial peace of the countryside and my state of mind than when I'm 'on the hay'. Make hay while the sun shines, runs the adage. But when will the sun shine here under the mountain in the west of England, where the average rainfall is

40 inches a year? I obsessively listen to weather forecasts on the radio, and nerdishly google weather reports on the internet, especially Scandinavian ones after our friend Annie convinces me they are the most reliable. Like their cars. I'm looking for the Volvo of forecasting.

And I find it. A Swedish forecaster who is right about almost every aspect of weather in south-west Herefordshire grants me ten whole dry days, starting 24 July.

I begin mowing at midday when the dew has dissipated, and the rising temperature is bringing up the pollen. The field smells like honey.

Funny, the things that take you back. A few years ago, I took the doors off the tractor cab, plus the back window, so sitting on the tractor was a cooler, more authentic experience. (And the de-pimping made it easier to bail out from; the electrics in the cab once went up in flames, and I had to jump for it; the dashboard still looks like something drawn by Dalí.) So every time I get in this almost open cab I am reminded of a photograph of my grandfather. He is driving a Ferguson T20, staring intently over his shoulder at the plough behind. The T20 is shiny, and is extra-shiny because of the rain. My grandfather is wrapped up in a grey raincoat. There is no cab.

I guess, because of my grandfather's age and the newness of the Fergie, that the photograph was taken in about 1958. They killed farming a year or so later. And they killed it by putting cabs on tractors. No longer was the farmer alive to the elements, or even close to the earth. All he did now was sit in a little mobile office, complete with heater and radio, pulling levers. I have sat in tractor cabs with air-con systems and plasma screen TVs, where you can literally put your feet up as the computer takes over.

Since a tractor cannot do a sharp double-back turn, hay-cutting, like ploughing, is done in elliptical 'lands', where one takes a wide turn at the end of the field, and eventually the parallel tracks of the tractor across the field meet.

After about twenty minutes of clickety scything by the bar mower, I have a good half-acre mown, the cut grass lying in long gorgeous Tudor braids, woven through with yellow from buttercups, red from clover, yellow-orange from bird's-foot trefoil, with a dash here and there of pink from ragged robin, and white from pignut, stitchwort and mouse-ear. The perfume from the cut sweet vernal is strong enough to drown the smell of the blue diesel smoke from the International's bonnet exhaust. And the sun is shining in a sky so pristinely blue it must be the first day of Genesis.

The bucolic reverie soon comes to a screeching, flailing end. The mower has hit a stone. (Yes, probably thrown up by my mouldywarps.) One of the cutters is bent, the other broken, the thing in bits.

Back at the house, I'm on the internet for an hour trying to find a replacement. The best deal I can find is £339 second-hand, but the snag is delivery: three days minimum. So I phone neighbours to see if I can borrow a mower, but the problem there is the mower either won't fit or it's in use. True enough. Through the open windows I can hear that the hills and the valley are alive with the sound of mowers straining and whirring.

I suppose I know what I am going to do, and I'm not even sure that, in some psychic manner, I did not arrange the wrecking stone.

In the cow byre, along with all the other implements and tack from a bygone age, is a scythe. A proper-job Grim Reaper's scythe, with a snake-curved hickory snath (shaft) and two grips. I will make hay by hand.

Funny, the things that take you back. I sharpen the scythe, handle to floor, with sideways-upwards flicks of the whetstone on the cutting edge. (More properly the edge is the 'beard'; there's a whole weird world of scythe terminology.) And I can see superimposed upon myself my father making D'Artagnan flicks with a steel to sharpen the Sunday carving knife.

The English scythe is a monster, with its heavy ash handle and coarse steel blade. Since this is an antique model, found in the byre when we moved in, it was made for an early-twentieth-century Herefordian, meaning someone about five foot six. But there is enough wormholed ash shaft for me to move the grips, which are bound to the shaft by metal clamps, by more than four inches. A good scythe, like a good shotgun, has to fit the individual. In that perfect world that I can never find, it should be bespoke.

The knack of scything is to keep the blade flat to the ground, so that it hovers a mere millimetre over the surface, and to swing the scythe around one's body in a circular arc. Knees should be bent, and the weight (presuming one is right-handed) transferred from the right leg to the left leg as one swings through. A man scything should be mistaken for a man performing tai chi. All this I know, because I have scythed before; as a teenager I used a scythe to mow the awkward areas between the trees in our orchard at home. Since then I have used a scythe to cut down weeds.

But when I mow grass this morning the grass mostly bends before the blade, then pops back up giggling. It does not help that the grass is drying by the minute; grass should be scythed when it is heavy with morning dew.

And the amount of labour required to keep swinging is fabulous. The blade needs to be sharpened ('honed' in the parlance) every ten minutes, and soon I am desperate for the scheduled stops to get the whetstone out from a bucket of water and flick it along the blade. My hands are also beginning to blister. My back aches, my face, in generous speak, 'has caught the sun'; in less kind words I look like a Brit at Benidorm. I have cut my finger deeply in a blazing piece of stupidity, by running it along the blade to check it was razored up. After two hours under the sun I have mown about a quarter of an acre. With a tractor this would take five minutes. If that.

Penny appears angelically out of the waterfall of perspiration with a mug of tea. 'How's it going?' she asks with a grimace.

'Fantastically!' I exclaim. Neither am I joking. Nothing in the last ten years of farming, with the exception of delivering calves, has given me such satisfaction.

I am in a state of near ecstasy. The grass I have actually mown lies in neat lines ('windrows') to the left of the blade's arc, and the sweet vernal, this close up, makes me think they must use it as deodorant in Arcadia. But it is the look and feel of the cut grass that is making me sing. Now that I have re-caught the knack of scything, the grass lies in fallen blades that silkily skitter and slide over each other, as though made from glass, not grass.

Hay from a mechanical mower is as much crushed as cut. Previously I had regarded this as a good thing, since bruised grass releases moisture more quickly and, after all, hay must dry. But this grass is a revelation. I can see and smell its quality.

That afternoon, I turn the windrows with a rake once.

I have searched the labyrinth of memory for a passage I once read about the virtue of hand-made hay, and finally found it in *The Worm Forgives the Plough*, John Stewart Collis's autobiographical account of working on the land in the Second World War:

> The agricultural labourer seldom praises anything, or admits that he enjoyed anything in the way of work; and none, save the old, object to the introduction of any mechanical device. But haymaking provided an exception to this – here at any rate. One and all, they not only hated the present job, they glorified the past. We made *hay* in they days; they said. It was regarded as a kind

of holiday time then, their families in the field, great picnics, not to mention lots of beer flowing.

In the evening I walk down the lanes between the rows of cut grass; there is something incongruous, almost exotic, about part of the familiar field being in braids. I am not the only one to appreciate the sense of holiday liberation, of the meadow becoming a new place. Under the far bank the rabbit young are racing in excited circles of astonishing speed, happy to be freed from the performance-decreasing constraint of tall grass.

The marshy, uncut corner amounting to an eighth of an acre is the remembrance of how things were only this morning, at dawn, on this side of the field. Here the devil's bit scabious, which likes the damp conditions, is in full flower. The devil's bit is a floral indicator of the antiquity of meadowland and, with its round, nodding lilac head, is more intensely beautiful than its name suggests.

Why the connection with Beelzebub? According to the fifteenth-century herbal *Hortus Sanitatis*, the plant was the Devil's very own dark material, the source of his power, and when the Virgin Mary put a stop to his evil, he bit off the root of the plant in annoyance.

Another, and contrary, explanation for the abruptly terminating root is that *Succisa pratensis* was so useful in healing human ailments that Lucifer, vexed, chewed its root to a stub to halve its efficacy. In this version of the plant's nomenclature, 'scabious' refers to its putative ability to cure skin diseases.

The devil's bit is certainly the food source for the marsh fritillary butterfly, one of Britain's rarer butterflies, whose numbers have declined by over 50 per cent in the last century. By July, the devil's bits in this corner are usually swarming with the black caterpillars of that species. Marsh fritillaries are whimsical insects, and local populations die out for no apparent reason. But ours will not die out, surely?

Next morning I am up with the skylark and down in the dewy meadow early. Today I am better prepared; I have found a pair of my father's old tan driving gloves, the sort that say 'Mine's a gin and tonic' and that my mother would have accessorized with her horsey headscarf from Country Casuals; and I have made, from a small plastic mineral-water bottle, a holder for my whetstone which can be hung off my belt. Since I do not have a proper hay rake, I have improved our garden rake by separating the wire tines with a pair of pliers.

After unpacking and spreading out the windrows with the rake, I begin scything. By elevenses, that immemorial time for the agricultural labourer to rest and slake his thirst, I have done three-quarters of an acre.

W

Hiss. Hiss. The gentle sound of grass falling to the blade.

Eight-foot-wide swathes of levelled grass lie in windrows in my wake. Ye olde yokel could probably mow this five-acre field in a day; inexperience means I will be doing well to cut an acre a day. Well, naivety and lack of fitness; modern men, even manual workers, cannot compete with a Victorian peasant, let alone a medieval one. But the Scandinavian weather god has granted me ten days, so I have time to spare.

The arc-action of swinging the blade and the repetitive swish as it razes the towering grass is entirely hypnotic and I fall into musing. The hay-cutter was always a field philosopher:

From a different line of work, my colleagues,
I bring you an idea. You smirk.
It's in the line of duty. Wipe off that smile, and
as our grandfathers used to say:

Ask the fellows who cut the hay.
From The Decade of Sheng Min, *translated by*
Ezra Pound

When people needed an oracle or an answer they always asked the hay man.

W

Robert Frost in 'Mowing' declared scything to be 'the sweetest dream that labor knows'. My long scythe too whispers to the earth and leaves the hay in rows.

I fell in love when I was fourteen with a flower meadow, perfectly set off by a wooden field gate beside the Wye. Almost all the things I love are to do with grass. Geese, sheep, cows, horses. Even dogs eat grass.

John Clare found his poems in a field. Sometimes I find words. There is nothing like working land for growing and reaping lines of prose.

W

I have never been so close to the animals of the meadow. To the leaf bug that grips grimly to the toppling blade of fescue, to the frog that decides to hop it, to the meadow brown butterfly that sups the nectar of the white clover even as the blade slithers towards it, to the rabbit lying in a form that does a white-tailed bolt.

I now know why the hay-cutters of yore tied up the bottom of their trousers before haying. A confused brown vole runs towards me and scurries up my leg, its tiny razor clees gripping into my flesh. As I am wearing voluminous shorts there is a nanosecond of some nervousness. My high-pitched shriek and passable attempt at reel dancing persuade the vole to jump off.

By one o'clock I have done about an acre. I spend the afternoon turning and drying the hay.

Despite all the seen and unseen motion of summer, there is a stillness in the landscape, as though it were trapped in a glass jar.

If anything, I am scything the next morning even earlier because by elevenses the heat is killing. I have now slipped into a rhythm of haymaking a Tudor peasant would understand. Cut in the morning; turn the hay in the afternoon. A steeping haze today takes away some of the fierceness of the incandescent sun, which is as welcome to the cut grass as it is for me. Trickily, the sun can be too bright for hay, bleaching it so it is as lifeless as shredded office paper.

The afternoon is also the time for the hay to be carted. Which, really, is where the fun begins. I have tied stock fencing around the sides of the 4-ton Weeks trailer so they have more height, and towed this down to the meadow behind the International. I have a pitchfork. So what could be easier than pitching the hay up into the trailer?

The answer comes the next afternoon. The amount of loose hay is wondrous. Forking up a ton of it on to the trailer massacres my back. I drive the tractor up to the yard, the trailer a Spanish galleon sailing behind me. I pitch the hay into the spare stables.

That evening I can barely move, and stoop around like a geriatric gnome with gastroenteritis.

After four days, I have mown about three acres, and have slowed to snail's pace. But I have had a brainwave. I have taken down to the meadow two 4.9 × 7.9m tarpaulins. When these are spread on the ground, the hay that has been raked into a 'haycock' mound can simply be rolled on to them. The tarpaulins can then be dragged behind the jeep back up to the farm, pulled to the right place, and the hay rolled off. Roll-on. Roll-off. No back-breaking pitchforking needed.

The only snag is that when I am towing the tarpaulins through the cow field the Red Poll attack them, butting at them with their heads. By standing on the second of the tarpaulins, they detach it, and start to eat the contents before I can shoo them away.

Their feasting is, I feel, a sort of compliment to my hay.

For the remainder of the day, they refuse to budge from the shade, where they repel insects with an African swish of their tails.

The horsefly is a silent and murderous biter. It comes not in ones or twos but whole battalions, drawn from afar by the smell of sweat. I become twitchy, like the horses do, expecting their attack, and begin slapping paranoically at any sensation on my skin. When the horseflies infiltrate past my defences and inject their probosces I smack them dead. *Haematopota pluvialis* means 'blood-drinker of the rain'. Half an inch long, slate grey, horseflies also have the nickname 'gadfly', a reference to either the fly's roving habits or the Middle English *gad* for pointed tool or iron (the same root as goad).

After crushing the biting flies, I wipe the blood off on my shirt. I look like the lead in *The Texas Chainsaw Massacre*.

For good measure I am also bitten by *Tabanus bovinus* (the horsefly with the hairy thorax) despite its supposed preference for cows.

In the evening I look upon my work. Eternity would not be enough to contain all the summer eves one could enjoy if they were like this, despite the horseflies. I chew on a stem of foxtail that escaped the hook.

Another blissful, rosy summer's eve, in Herefordshire, the unknown county.

The buzzard young are on the wing, making the meadow their nursery hunting ground. A wood pigeon coos drowsily in the oaks. Grasshoppers are chirping . . . For a second I think I hear a nightjar.

There was a nightjar here once, on a similarly warm and windless night. In the near dark, its churring seemed to emerge from the very landscape, as if the earth was vibrating. Then, for a second, the bird flew up against the sinking sun and performed a silhouetted cartwheel along the mountain top. Or so it appeared in a glorious moment of trompe l'oeil. Then it flew away. For the nightjar, the field was just a stopover on the way to Ewyas Harold Common or on to the mountain. It was not home.

I have seen them on the common in daylight, reposing like lizards in the trees. The nightjar is not an attractive bird, what with its drab plumage and gaping mouth; it was said to steal the milk from goats, hence the local name of goatsucker. But the nightjar is entirely an insect eater, catching its prey on the wing like the martins, only at night, using the bristle to the sides of its mouth to funnel insects into the chasm of no return.

I have decided to sleep under the stars.

They say that if you can remember the sixties you weren't really there. By the same token, if your night under the stars was free of insect bites and rustling you didn't really do it. But better to lie out in a sleeping bag in the open than in a tent, which is only another form of house. Tonight heaven is my roof, the hedges my walls. A procession of late night birds and early night bats fly over me, and a hedgehog snuffles along to give me a fright with a human-sounding snort.

The jackdaws do not seem to sleep, but jackdaw young have no sense of danger, so need great and loud teaching. Lots of it, apparently.

Then the field folds me in soft wings.

One man went to mow, went to mow a meadow.

Haying, of course, used to be a communal activity, although the raciness of traditional English folk songs about mowing suggests that the day's activity extended to rolls in the stuff:

> *'Twas in the merry month of May in the Springtime*
> *of the year,*

All down in yonder meadows there runs a river clear,
And to see those little fishes how they do sport and play,
Caused many a lad and many a lass to go there a-making hay.

In comes three jolly scythesmen to cut those meadows down,
With a good leathern bottle and the ale that is so brown;
For there's many a smart young labouring man comes here his skill to try,
He whets, he mows, and he stoutly blows for the grass cuts devilish dry.

Then in come both Will and Tom with pitchfork and with rake,
And likewise black-eyed Susan the hay all for to make;
For the sun did shine most glorious and the small birds they did sing,
From the morning till the evening as we goes hay-making.

Then just as bright Phoebus the sun was a-going down,

Along comes two merry piping men approaching from the town.
They pulled out the tabor and pipes, which made the hay-making girls to sing,
They all threw down their forks and rakes and left off haymaking.

They called for a dance and they jigged it along,
They all lay on the haycocks till the rising of the sun.
With 'jug! jug! jug! and sweet jug!' how the nightingale did sing!
From the evening till the morning as we goes hay-making.

Another song, 'As I Was a-Walking', recalled:

A brisk young sailor walked the field
To see all the pleasures Flora would yield
He saw a maid dressed in a smock
Busy a-raking, all round the hay-cock.

When it is really hot I, to the children's ire, put on a terrible Wild West voice and say, 'It's hotter'n hell.'

I have to get up at dawn to beat the heat, but even so, by ten it is warmer than Hades. This fierce morning light shows every detail of the mountain: each sheep, each scattered hawthorn tree, the tumbled careworn rock face of Red Darren.

More and more often I rest by the river. Spit bubbles come towards me in flotillas. Leaves roll along. The kingfisher flies past. 'Zeet. Zeet'. The mallard mother softly calls to her four remaining chicks, now as camouflaged brown as she, and they paddle away in a safety routine now familiar to all. Walking back into the meadow I alarm a sunning gatekeeper which flutters up and off; the golden butterfly takes its name exactly from this habit of rising up, which reminded people in past centuries of the men employed to mind tolls, who would sit up when customers appeared.

It is late July, and this is the first gatekeeper I have noted this year.

The adults lay their eggs singly on grasses beneath bramble, blackthorn and hawthorn bushes where the grass stems are not grazed by animals. At night I take a torch to a corner of Marsh Field protected by a barbed wire fence. Gatekeeper caterpillars, which are nocturnal, are a nondescript brown, and feed on grasses, with a preference for fine species such as the fescues, bents and meadow grasses. Despite an hour-long search I find no gatekeeper caterpillars.

But the devil's bit scabious in the meadow is hanging with the mace-spiked black caterpillars of the marsh fritillary.

And meadowsweet now trips lightly out of the hedge and into the damp ground by the newt ditch. A common sight in the old British countryside, *Filipendula ulmaria* has disappeared along with the water meadows that were its main abode. The creamy summer flowers are sweet to smell when rubbed under my nose, though an oenologist would note the hints of almond. In Tudor Britain meadowsweet was one of the principal 'strewing herbs', scattered on floors as an air-freshener, and the great herbalist Gerard went so far as to say that meadowsweet

> farre excelle all other strowing herbs for to decke up houses, to strowe in chambers, halls and banqueting-houses in the summer-time, for the smell thereof makes the heart merrie and joyful and delighteth the senses. Neither doth it cause headache, or loathsomeness to meet, as some other sweete smelling herbs do.

It was meadowsweet, 'and willows, willow-herb, and grass' that caught Edward Thomas's eye on that baking day of June 1914 when his steam train drew up unwontedly at Adlestrop station, and he made a photograph in poetry of pastoral England on the eve of war.

Some say the flowers of meadowsweet fizz, others say the blossom is spun as finely as vanilla candyfloss. All I know is that the feminine delicacy of the meadowsweet's flowers, which are in bloom from June to September, is hinted at by the names lady of the meadow, maids of the meadow and queen of the meadow. In the Welsh mythical tales collected in the *Mabinogion*, the magicians Math and Gwydion take flowers of oak, of broom and of meadowsweet to create 'the fairest and most beautiful maiden anyone had ever seen', Blodeuwedd, or 'Flower Face'. So womanishly graceful is meadowsweet that in Irish mythology, Cú Chulainn, the testosteroned hero of the Ulster Cycle, used meadowsweet baths to calm his rages.

Like the fair Blodeuwedd, who would turn out to be a homicidal adulteress, meadowsweet has a secret. The dark leaves whiff of childhood TCP. So the plant's contradictory nature is also caught in such oxymoronic local names as bittersweet and courtship-and-matrimony. Even meadowsweet is misleading, because the plant is not named for its liking for meadows but for the role its leaves played in embittering and aromatizing medieval mead. Hence its appearance as Middle English *medewurte* in Chaucer's *The Knight's Tale*. Meadowsweet, it might be said, is a plant with a history as well as literature. Evidence of meadowsweet has been found in Bronze Age burial sites, and the Druids are said to have ranked it as one of their most sacred herbs. When Gerard noted the plant's characteristic of not causing headaches he was more right than he possibly knew; the flower head contains salicylic acid, from which, in 1897, Felix Hoffmann created a synthetically altered version of salicin. The new drug was named aspirin by Hoffmann's employer, Bayer AG, after the old botanical name for meadowsweet, *Spiraea ulmaria*. This gave rise to the class of drugs known as non-steroidal anti-inflammatory drugs (NSAIDs).

On this furnace-hot afternoon, when no bird can be bothered to sing, and I am unsure whether the metre-tall meadowsweet looks more like debutantes gathered for a ball or a cresting white wave, I have picked twenty full sprays of meadowsweet for cottage country wine. So deadly intoxicating is the nectar that hosts of hoverflies will hardly let go their sucking tongues even as I put the flower heads in the carrier. Meadowsweet is beloved of insects; the gall midge *Dasineura pustulans* has already burrowed into the leaves and left unseemly yellow blisters; and the plant is the larval food for all manner of butterflies and moths.

Moths might be the plain cousins of butterflies, but surely they have the poetry in their names? Among the moths whose young feed on meadowsweet alone are lesser cream wave (how suitable a name for a meadowsweet-loving moth is that!), bilberry tortrix, glaucous shears, Hebrew character, least yellow underwing, scarce vapourer. And who wouldn't want to see a powdered quaker? But it is the rather prosaic brown-spot pinion that is the commonest sucker of Lower Meadow's meadowsweet.

Late that night I take the dogs down with me to check on the cows in Marsh Field. The Labradors rush through the gate into the meadow and away, racing to where the meadowsweet stands illuminated. A little owl rises up from the hedge top with a shriek of annoyance. An owl in Welsh is *blodeuwedd*, because Flower Face's punishment for her crimes was that she could never show her face again in the light of day.

The very whiteness of meadowsweet is a beacon to moths. My torch beam shows fox moths nectaring and one which I think is a satyr pug, another splendid name.

AUGUST

Rabbit

1 AUGUST LAMMAS DAY, from the Saxon *Leffmesseday*, meaning loaf mass. This is the traditional day for cattle to be returned to the hay field to graze the aftermath of hay mowing. Not in this field, not this year, for I am still scything away, still turning and hauling.

It has become a battle. I am not sure whether I will finish the field, or it will finish me. As an alternative to scything, I bring down the brush-cutter, a heavy-duty strimmer.

Something about strimming makes me tighten my jaw in concentration. The brush-cutter does well enough, except it does not leave the cut grass in windrows but all scattered, and with double cuts that make the hay more like fluffy chaff.

And in the noise and fumes I lose the peacefulness of it all. I do a morning's worth of mowing with the brush-cutter, then turn the damned thing off, to listen to the sound of one moment:

Breeze.

Huzz.

Peow.

Buzz.

And the cautious pitter of the Escley, where it

steps down a foot from one rock bed to the next. Not a car, not a plane, not an internal combustion engine. White castles of clouds make a stately procession through the sky.

I'm not sure of the exact time; I don't wear a watch. Farming isn't a by-the-clock job; it's a job determined by light and weather. And anyway the thistles act as good enough sundials; they are casting little shadow, the sun is overhead. Around midday then.

Heat and dust. The meadow still to be scythed is running and hopping with butterflies. A wood pigeon is mesmerically cooing from deep in the ash, its mate sitting on the rickety raft of twigs that passes for a nest. This is the second brood of the year. Where the first grass was cut is a parched shade of brown, save for where two goldfinches are pecking inquisitively, gleaning the fallen heads of grass and flower seeds, a strangely peasant act for such a courtier of a bird. They have two young with them, brown nondescript balls, not yet clothed in gold-and-scarlet finery.

So, it's back to grim reaping and aching tranquillity. Long days and red-rimmed eyes, a face-mask of grey pollen. And a stupid peaked-cap to wear, bought on a family holiday in the Dordogne, emblazoned with the advert 'Camping Soleil Plage'. I hit a patch of meadow with an abundance of sweet vernal grass, always distinctive because of its cylindrical flower spikes and the vanilla incense it emits, and which almost transports me into sleep. It would be the sleep of Caliban. When I waked I would cry to dream again.

Unfortunately, I have already run out of space to store the hay. Loose hay takes up an improbable amount of room. I have filled the stables, and most of the cow byre.

By my estimate I am getting off more than 1.25 tons of grass per acre. I still have three acres' worth to house.

I am going to have to build a hayrick.

I also have another field to mow, the six-acre Road Field. There is no physical possibility of my doing it by hand, so I phone Roy Phillips, the agricultural contractor. The children are delighted, since Roy will turn the grass into big round silage bales, with modern black plastic wrapping.

I resume field philosophy: the green sameness of modern grassland could be a metaphor for rural life. Where are the characters in the village? As late as thirty years ago, when I caught the village bus home from school on market day it would bulge with old women in tweed coats with cardboard boxes containing indignant chickens on their laps. The aroma of vintage cloth, medicated soap (all country people used medicated soap until the 1980s) and chicken excrement was thick and unforgettable. The chickens had been bought from the livestock market in the centre of

Hereford. The livestock market has now been relocated in favour of a shopping centre. And there is another metaphor.

Sometimes Julie, the daughter of Mr Preece, would sit on the carpety seat next to me. Mr Preece – never without turned-down wellington tops and red braces – ran the smallholding at Woolhope from which we always obtained our Christmas turkey. I was never sure whether it was Julie's boss eye that I disliked, or that I somehow got her mentally confused with the turkey. There is not much glamour in poultry farming.

In a photograph of two of my aunts when young, they wear white dresses and white ankle socks. Presumably it is their confirmation day. This is some time in the 1940s, when my grandfather was a farm manager for the Prudential. Behind Josephine and Madeleine is a row of tall detached houses end on.

It requires a second look to see that the houses are actually perfectly constructed and spaced haystacks.

When I went to bed, my haystack (at a mere fifteen feet in height) was fine and upstanding, and had a quite fetching pitched top. By morning it has taken a distinctly Pisa-ish lean. I am pondering what to do when I hear a hellish rattling and a white DAF van appears over the brow of the track down to the farm and eases forward in slow motion. There is a tatty brown rope around the middle of the van, like a belt, to keep the side doors on. I have not seen this van since we moved here from Abbeydore, six miles away, more than half a decade ago.

Geoff Bridger is known to all as 'the black-and-tan man' because his black-and-tan Jack Russell bitch constantly goes missing, leaving Geoff to knock on neighbourhood doors to ask, "ave you seen a little black-and-tan dog?'

But he has never had to search this far for her before. The van coughs to a halt on the yard beside me; the driver's door shoots back. A face with just two teeth, the bottom canines, peers out.

"ave you seen a little black-and-tan dog?'

Geoff squints at me harder, pushing his head further out from his stained check shirt.

'Oh, it's you. So this is where you've been hiding yourselves.'

Geoff fell out with our friends, Nick and Alice, when he was doing some garden digging for them; Nick accused him of working too slowly. Geoff took umbrage and decided to fall out with us too.

'Well, if you see 'er maybe you could bring 'er back. If it's not too much bother,' he adds.

As Geoff begins to close the van door he catches sight of the hayrick.

'Oh, what the hell you wanting to do that for?'

Before I can say anything, Geoff is out of the van walking around the stack. He scrutinizes the bottom.

'Well, you've got it up on pallets, that's good and dry.'

Geoff has a strange brightness in his eye. 'I haven't seen a haystack for' – he shakes his head grimly – 'what? Forty years?

'C'mon,' he says, 'let's get the bugger propped up before 'e slides away.'

With some old railway sleepers and a couple of field gates, the stack is propped up.

'If anyone says anything, tell 'em you lost yer watch and had to take the side of the stack away. That's what we always used to say when it went arse over.'

Getting back in his truck (a van is always a truck in Hereford) he says, 'Good hay though, that.'

Which, from one of the last country characters, is praise indeed.

I shake his hand, and hope he finds his dog.

My mistake in rick building is thrown into cruel and sharp relief by the success of another tyro, John Stewart Collis, down on the farm in the 1940s:

> The thing in building is to get your walls up straight, which I found easier to understand than to do, since there is a strong psychological feeling against putting the hay out – one always feels it will fall over, not realising how strongly it

will be bound by the hay that goes behind it (for hay binds like brambles, as you find quick enough when you try to take it out). This tendency against the perpendicular is most strong at the corners when it is most necessary to oppose it and be bold. The great thing, I found, was to put two helpings at the corners, and not be faced with the psychologically distressing sight of a sloping margin. However, it all went well, and I roofed it in the approved Gothic style. It needed no props – and, I believe it who will, 'E was heard to say – 'One of the best ricks we've done.' It was on the highest level of the field, and so as we went away in the evening when it was getting dark, it looked wonderful, to me, against the sky – all those untidy bundles that I had been dealing with throughout the day now compressed into a pure solid, the pointed roof traced blackly and with geometrical straightness and sharpness against the light. Going away from it, down the sloping field with the others, I tried not to turn my head too often to have a look at it.

3 AUGUST The last waving, dew-glittering pool of grass to be scythed. Close to the thicket, the understorey of the grass has been tunnelled along by short-tailed voles

(*Microtus agrestis*); the blade of the scythe exposes their hidden, much trodden runways to the light. Ahead of me the voles flee, squeaking, a petty plague. The heat of summer has boomed the voles' numbers; young are weaned after fourteen days, and females may have four broods in the middle months of the year. The blade decapitates a doughnut of finely woven grass; inside are four naked vole babies. I cover them up.

There are 80 million field voles in Britain, a grey and unassuming mammal that is the meal for almost every land predator. I can almost feel the eyes of foxes and raptors fixing on the scene.

Across in the grass islands left for the pipits young swallows skim the meadow, learning the art of the chase. The blackbirds work the margins of the field, staying close to the hedge; they are moulting and their flight is hampered. The hedge is safety; a buzzard drifts over on the breeze, and the fearful blackbirds flutter to the hedge, perching with their trademark steadying tilt of the tail.

Clouds creep over the field in the afternoon, threatening rain.

4 AUGUST After nine days I am done, the hay-cutting is finished. Four and a half acres are as smooth as baize, with two long-haired islands floating in them. I

think I can genuinely say I know every blade of grass in this meadow.

Roy Phillips has cut and baled the Road Field, and thirty or more black sheathed bales are lying around. Freda and Tristram are delighted. They like jumping and climbing on the bales; there is something of the playground or school gym about a round bale. Best of all, as far as my son is concerned, is that I will allow him to move the bales down from Road Field when needed by pushing them along with the jeep, in a sort of automobile football dribble.

5 AUGUST I am not sure whether I am inspired by Collis, or thrown into competition across the decades. With the last thirty hauls of hay, I build another rick; it is hardly a skyscraper, but for a bungalow it is, I venture, rather neat.

7 AUGUST The day blows up black. House martins spark over the field. The rain comes swarming over the mountain from the west. (The Scandinavian forecaster was correct to the day.)

According to folklore, the purple foxglove is so called because Reynard dons its flowers on his feet, so as to be able to creep in magical silence up to hens and nab them. Perhaps the fox which stole the fluorescent white chicken used such guile. Lying slap in the middle of the meadow is a yobbish scattering of feathers, and the hollowed-out corpse of a fowl. For one mad moment I think someone has put a joke-shop rubber chicken in the field.

No, on close acquaintance the body is real. I leave it for the rain and buzzards.

The Herefordshire name for foxglove is bloody man's fingers. The plant grows haphazardly along the exsiccated back bank of the Grove ditch.

The chicken is not ours but our neighbour's. Our chickens are surrounded by an electric fence, which is a reasonable although not infallible deterrent. The foxes yelp when their wet noses touch it.

There is something more the foxes may have in their memory, genetically passed down the generations. I am an Old Testament poultry-keeper. I say a life for a life, and have a gun that speaks death.

And how many of the mallard young have cheated the fox's snare-trap jaws? I have not seen the wild ducks for a fortnight or more.

12 AUGUST The evening is 'close', humid. Down in the meadow, where I take a rest from trimming sheep's

feet, a can of Ruddles for reward, I touch the aftermath pea-green grass. It is fine, tender, new.

A stertorous old man disturbs my peace. Well, for one second at least, I believed it to be an old man.

Hedgehogs are the quintessential mammals of the hedgerow, fossicking around the bottom for slugs, beetles and other invertebrates, and using its secret dark heart for hibernation in winter. In summertime, they will sometimes make a daytime nest in the hedge bottom for a snooze. Although resident in Britain since the last ice age, the hedgehog has declined by as much as 25 per cent in ten years, according to surveys by the People's Trust for Endangered Species. They wander as much as a mile and a half a night.

It is not an exaggeration that hedgehogs snore; he is wheezing away under the brambles in my broken-down wild corner.

That night: a veil of cloud keeps the huge moon at bay, which retorts by dousing the cloud with an oil-on-water rainbow sheen. The tawny owls are hunting over the meadow, like giant, dimly perceived moths. July and August are the months of the second diaspora of mouldywarp young, when they travel the great distance of five hundred yards or so from their birth burrow. One crawls childishly along in my torch beam, the locomotive power almost entirely provided by the back legs. The long nose sniffs ceaselessly for danger.

W

13 AUGUST A typical sultry August 'dog-day', named for Sirius, the Dog Star, which now rises and sets with the sun. I see the first brimstone butterfly of the year, a flitting shock of sulphur in and out of the hazel hedge. Brimstones are one of the few butterflies that pass the winter as an adult; they emerge from their pupa cabinet at the end of July. She flutters up high, and an unnoticed male launches after her. They transport themselves on a breath of air to the shady thicket, where she dives down, the male (and I) in hot pursuit. On the ivy wrapped around an alder she disports herself, and there she mates.

Coition was still in progress the next afternoon.

W

14 AUGUST An anaconda of fog comes slithering up the narrow valley bottom in the evening, on, on, following precisely every turn of the river, to lie suffocatingly on top of us.

A herd of weaned calves have been put in the field next to the paddock. Our girls sing to them, heads flat out, a tuneful lowing.

W

17 AUGUST Penny and I have a sort of busman's day off, and go for a walk around The Parks, the wild-flower nature reserve at Dulas, three miles away. For about five hundred years The Parks used to belong to Dulas Court, the home of the Dulas branch of the Parry family, until they sold it in 1840. The new owners demolished the old house and chapel, and built afresh. Clearly the Fieldens did not want the great Victorian unwashed within sniffing distance, and relocated the chapel in a field across the lane. This turned out to be a favour to botany; the grassland between the graves is the finest species-rich grassland in the valley, and home to the common spotted orchid. God's acre is one of the few places safe from HS2 and Bovis.

The Parks is always a bit of a revelation. Lower Meadow is a fairly traditionally maintained hay field, whereas The Parks is, as we say in Herefordshire, 'the proper job'. The flowers seem to outnumber the grasses.

18 AUGUST In the newt ditch I spot baby palmate newts, an inch long, gills like fins. Chameleon frogs too, water boatmen, pond skaters. The air whines with mosquitoes; their name comes from the Greek *muia*, an onomatopoeic rendering of the noise they make when flying.

The grass in the uncut islands is so bronzed it looks like ripe wheat. In the belt of thistles the heads are white and ripe and explode at the touch, unless you are very gentle; they could be used as make-up brushes for a woman's skin. There's a languor on the scene; a peacock butterfly dries its wings on hogweed by the river, much as a dog stretches out in self-satisfaction before a fire.

The badgers like the green baize of my mown areas; three of them appear late, around midnight, and waddle along slurping up worms in the moonlight.

W

22 AUGUST Evening. The hills and mountain are smoked white by haze. I am standing in the field with my shotgun, eyeing the rabbits. There are about twelve of them, mostly grazing, one or two cleaning their faces with their paws. One of the first to come out, a heavy buck, has defecated on top of the ant mound closest to the warren, a signal of territorial intent.

I am unsure about the current kinship pattern in the warren, or its relation to other warrens nearby. There is quite a lot of rabbit fur around the anthills, the residue of rabbit in-fighting. In a Gerald Ratner moment Richard Adams, the author of *Watership Down*, declared that real rabbits were 'boring'. Animals have what I term a 'danger diameter'; the rabbits consider that at forty yards I am too far away to kill. Animals can also detect motive; as soon as I level the shotgun, they will be alert, then they will scram.

Playing God with a gun is not always fun; I settle on a luckless young rabbit (but not a baby) nearest me, push the safety catch off, advance five yards and fire. Whether it has been with a bow and arrow, a falcon or a gun, killing game in a meadow is as old as the meadows.

27 AUGUST Note on paper, scrawled during the day. 'Squirrel in hazels pulling the trees apart to get at the not yet ripe nuts. Not very eco.' I should have added that the chandeliers of elderberries are almost blackly ripe. Plants are a calendar marking the days, the seasons.

In the afternoon, driving down the lane, I get a fright when a polecat pokes its head out of the ditch to give me a malevolent look. I stop the car. Through the open window the rosebay willow herbs tremble under the weight of the sucking bees, and a hidden yellowhammer raps out its hard call of 'little bit of bread and no cheese'. For a minute the ferret-faced polecat and I are lost in mutual suspicion. There is no need for me to observe the Jefferies rule of observation; at two feet in length the polecat is all too aware of its own power. I tire first, and drive off. In the rear-view mirror I see the polecat staring after me. Despite the heat, I shiver.

29 AUGUST The distance is veiled in amaranth vapour; the mountain heather an inferno. After a day in the tractor topping weeds on some land we rent about six miles away, I lie for five minutes' bliss in the field. I see a fox (one of the young from the den) eat a blackberry in Marsh Field, by standing on its legs and plucking with its mouth. A lovely moment for a lazy man.

30 AUGUST A robin is singing in the copse. Past the August moult, it is laying early if wistful claim to a winter territory. A great tit provides accompaniment, with its 'teacher, teacher' refrain. There's a tiredness in the air.

SEPTEMBER

Damselfly

THE MONTH OF half summer, half autumn. Half sophistication, half barbarity.

I have put the cows in the simmering field. Under the shade of the Grove hazels, only their candy Ermintrude noses betray them. When the Red Poll deign to advance into the meadow to graze on the aftermath, their coats shine like conkers. Cattle are good for flowery meadows because their grazing creates a variety of sward heights, important for providing suitable nesting and feeding conditions for birds.

Despite the lingering tones of summer, you can tell autumn is on its way. The thistles and nettles along the Marsh Field hedge are bowing over, elderly and huddled, unable to support their own weight.

A crow rows through the sky.

Wasps soporifically suck on blackberries.

3 SEPTEMBER I suddenly realize that the swifts have gone. No fanfare. Just a prestidigitator's trick, a disappearance into the morning's mist. Inside I

sigh a little. One of life's allotment of summers is over.

As I take the morning walk to check the sheep, three mallard whistle down out of the sky and alight, skidding on the river; I stalk them later with binoculars. I am near certain it is the mallard mother with two of her offspring; the pair have the grey beak of youth.

W

6 SEPTEMBER An azure damselfly in the meadow, a slender jewel kept aloft on the gauze wings of fairies. Damselflies, along with their cousins the dragonflies, make up the scientific order Odonata and are almost unchanged since prehistoric times. They are marvels of engineering, able to alter the angle and beat of each of their four wings so they can fly up, down, sideways or backwards and hover for up to a minute. Some dragonflies can reach speeds in excess of 30mph. Voracious predators, the adults locate their flying meat meal by use of their bulbous outsize eyes, which can see in almost all directions at once. They will eat almost any insect with wings, even bees. And crane flies.

Masses of crane flies or daddy-long-legs are now hatching in the field, their jerky, marionette motion a constant loathsome intrusion. One flies into my face, its limbs trailing across my cheek. I'd like to brush it away, but know the movement would scare the adolescent fox across the field. Her coat has lost its baby grey, and she is a resplendent starlet in her red fur.

She is sitting, transfixed.

A female spotted flycatcher (a notoriously late nester) is repetitively launching herself off the top strand of barbed wire under the copse, seizing a gangling crane fly from the air before returning to her vantage point on the fence. She is a shooting sylph of silver. Her solitary offspring, also sitting between the barbs, is Bunteresque, with a silent, insistent sense of entitlement.

The teenage vixen moves closer to the flycatcher, and sits again.

I have presumed too evil an intention to the fox. She leaps not after a flycatcher but after a crane fly. For over twenty minutes of this sultry, languid evening fox and flycatcher leap at crane flies side by side.

Crane flies, the hatched adult of the leatherjacket, are not an entirely satisfactory meal for a fox. The vixen realizes this, and slopes off.

Crane flies are feast enough for a flycatcher. The next day, fuelled up, it flies south for winter. The youngster also disappears. Of the summer migrants only the chiffchaff, blackcap, house martins and swallows now remain.

*

Fish-scales of yellow moonlight fall on the river. The hedges have the unmistakable liquorice smell of dying leaves that signals autumn.

Cows might be domesticated, but wild habits die hard. They lie tonight in a circle, facing out, to see the danger that might come from any direction. Something about the angle of the moonlight increases both their antiquity and size. They might be mastodons, the black beasts I am crouching behind, hiding.

I have come down to the meadow, alarmed at the unholy noise in the night. But the cattle are quiet; the villain of the peace is twenty yards away on the silver sward. The old boar badger is in the meadow, making the throaty gargle that passes for wooing in the *Meles meles* species.

After an agony of cramp for me, the dominant sow, who has been his wife for two years now, deigns to make an appearance, coquettishly slinking under the wire.

A few moments of half-hearted kiss-chase follow; badgers, though, are not big on foreplay. Grunting furiously, the boar clambers on the back of the sow, clamps her neck between his teeth to keep her in place. The light is too indistinct to make out more; my mind, for reasons not difficult to fathom, recollects that Victorian fathers of the bride gave tie pins made of badger penis-bone to the new son-in-law to ensure fertility.

It is not the sight or sound of badgers mating that is remarkable. It is the cloud of rank musk, overpowering at even twenty yards downwind, that the act unleashes. If you have smelt it you will know why badgers are related to skunks.

W

10 SEPTEMBER Along the luxuriant river bank, underneath the alders, it could be the green time of spring; one has to look hard for the signals of autumn. They are there, though: the burs from the burdock which stick to the dog, the empty seed envelopes of the hemlock, and the solitary willow warbler poking about perplexedly in the goat willow. (Of course.) The willow warbler is passing through; but then we are all passing through.

When I walk up the bank into the field the meadow pipit fledgling is making tentative ascents into the golden air. I do not know what became of its siblings.

Bizarrely, a skylark launches up too, as if to give a flight masterclass.

Yesterday's rain and today's warmth have brought the slugs out; a pair of black slugs (*Arion ater*) circle and lick each other, their mucus-covered bodies as moist as the green grass on which they lie. Slugs are hermaphrodite; entwined, each releases its penis, a white tentacle which clasps the penis of the other.

W

14 SEPTEMBER This silent void is a shock. No evening chorus, no universal hum of insects, and the lambs, now grown up, no longer bleat, they only eat, heads permanently fixed to the ground. Children's plastic farm toys would be more real.

The chiffchaff has flown.

W

17 SEPTEMBER I move the cattle out of the field and into a pen for their annual, government-ordered tuberculosis test. They smell divine, of buttercups, of vernal, of grasses galore, as good cows should. Sunshine glints off their backs, full of summer bloom.

This is my least favourite day of the year. Lined up in the race, a narrow metal-railed corridor, the cows are injected, by the vet, in the neck with an instrument that looks suspiciously like a staple gun. If they test positive to the reagent they are to be slaughtered. No ifs, no buts. I have to wait four days for the vet, dressed top to toe in green plastic and rubber, to return and pronounce.

The days of James Herriot, of tweed and a bacon

sandwich over a cup of tea in a farmhouse kitchen, are long gone. Vets today dress like forensic scientists at the scene of a crime.

W

20 SEPTEMBER I watch a ladybird climbing a skyscraper of grass. *Propylaea punctata*: a perfect black-and-yellow chequerboard that should have been designed by Issigonis in the 1960s.

Underneath the hedge a family of pied wagtails runs through the glistening grass. The moment of minimalist art is improved by two equally black-and-white magpies strutting through the sward. And the hedgerow is burgeoning with red and purple: rose hips, elderberries, sloes, blackberries, honeysuckle, and the luxurious forbidden berries of bryony and deadly nightshade.

A magpie halts to look quizzically at the jumble of grey feathers below the oak tree; some predator has killed and plucked the two wood pigeon squabs.

In the pen, I feed the cows hay and cattle cake; while they are otherwise engaged I run my hands over their thick, muscular necks to feel for a reactive lump.

One has a swelling.

W

23 SEPTEMBER A morning in which the field is garrotted by mist. This moist warm weather brings on the autumn flush of grass, and the mushrooms; in the sward there are bronze turf mottlegill mushrooms and rare yellow waxcaps (*Hygrocybe chlorophana*). Growing out of the cow pats are liberty caps.

The ivy on the elder is in bloom, though its unostentatious, spherical greeny-yellow flowers hardly qualify, I feel, as flowers in the visual sense. Nonetheless they are an important source of autumn nectar for the last bees, moths and butterflies of the year.

Some juvenile house martins are still here. The adults went yesterday. From the dead elm in Bank Field a great spotted woodpecker 'tschicks' territorially.

This is all happening somewhere far off, as though I am looking at a hushed English pastoral scene down the wrong end of a telescope. The cows are back in the race waiting. They are listless; the dominant cows in the herd order, Margot and her daughter, Mirabelle, are butting those in front, and the metal barriers are starting to screech alarmingly under the pressure. This is my fault; the cattle are sensitive to my anxiety. I am so wired up about the TB check, I can hardly breathe.

The veterinary surgery rings. The vet, Will Jacobs, is going to be late. Some time about three o'clock Jacobs comes wearily down the track. The usual clambering into the protective suit. His hand running over each cow's neck. All good until he gets to Melissa, Melissa with the bump. Out come the measuring callipers.

I have been here before. Everything depends on the size of the bump. Measured once. Measured twice. Measured three times.

'She's just inside the limit.' I could jump for joy. 'They're all yours for another year.'

I let the cows out. We run around together.

25 SEPTEMBER Amid the dew-laden webs of millions of *Linyphiid* spiders, which medieval shepherds believed caused braxy (a digestive disorder of sheep), a grey squirrel is picking up hazel nuts. There is a palpable urgency to his or her action. Although grey squirrels do not hibernate they need to larder up against hard times.

At the end of the day: on this night of a waxing moon my shadow is giant across the field.

Another busman's day out. We drive over in the drizzle to Turnastone Court at Vowchurch, six miles away in the broad, flat Golden Valley. This is the Parry heartland. There is an effigy of Blanche herself in Bacton church, on a hill that always seems to be in cold shadow. The effigy has a minor footnote in history because it incorporates the first image of Elizabeth I as Gloriana. Of more interest to me: Blanche looks remarkably like my grandmother.

The meadows at Turnastone are the remnants of a utopian agricultural project set up beside the Dore river by Rowland Vaughan, who wrote a book published in 1610 on 'his Most Approved and Long experienced Water Workes containing the manner of Winter and Summer drowning'.

Vaughan is usually credited with the invention of downward-floated water meadows or bedworks, although some scholarship suggests he merely developed an extant system. (The field name 'le Flote' at Kimbolton in Herefordshire pre-dates Vaughan's book by a generous historical margin.) To an extent, this is pharisaical stuff; what Vaughan did and popularized was the temporary flooding of grassland via water diversion, channels and sluice gates.

Vaughan was the great-nephew of Blanche Parry. (He complained that his spirit was too tender to endure the 'bitterness' of Dame Blanche's 'humor' and that he was forced by her 'crabbed authority' to fight in the Irish war.) The Parrys and the Vaughans had intermarried for at least a hundred years, and Rowland did not buck the habit. He married the Parry girl who inherited the main family estate, Newcourt, giving him ownership of all the land on the west bank of the Dore from Peterchurch to Bacton.

Famously, the idea for 'drowning' grass came to Vaughan when he was walking his estate to check on the miller (a notably shifty species). As he strode along he noticed that a mole had burrowed into the millstream bank, and where the water oozed out through the molehill the grass was luscious.

Vaughan spent twenty years constructing an irrigation system in the Golden Valley – approximately from 1584 to 1604 – whereby his grass could be flooded to promote its growth. His main artificial channel was the three-mile-long Trench Royal, which diverted water from the Dore on to the fields, then away again via a sluice gate. The use of flooding increased the yearly value of the land from £40 to £300 per year.

Although many thought Vaughan mad, his method was demonstrably successful and attracted great acclaim. A 'panegyricke' written by the poet John Davies praised Vaughan's drownings of meadows in effusive rural imagery:

His royall TRENCH (that all the rest commands
And holds the Sperme of Herbage by a Spring)
Infuseth in the wombe of sterile Lands,
The Liquid seede that makes them Plenty bring.

Here, two of the inferior Elements
(Joyning in Coïtu) Water on the Leaze

(Like Sperme most active in such complements)
Begets the full-panche Foison of Increase:

For, through Earths rifts into her hollow wombe,
(Where Nature doth her Twyning-Issue frame)
The water soakes, whereof doth kindly come
Full-Barnes, to joy the Lords that hold the same:

For, as all Womens wombes do barren seeme,
That never had societie of Men;
So fertill Grounds we often barren deeme,
Whose Bowells, Water fills not now and then.

Mind you, John Davies was a kinsman.

Six years later, in 1610, Rowland published his book describing the system. In it he claimed that the Trench Royal was navigable, and was being used to ship goods from one end of the estate to the other. The book also claimed that he established an ideal community for two thousand workers, who were all decked out with fetching scarlet caps.

Water meadows became fairly common in Herefordshire. The temporary diversion of water (ideally an inch deep) over grassland in winter encouraged the growth of grass before the growing season and provided stock with an early 'bite'. In some cases, a further period of irrigation allowed a second or even third hay crop to be taken. Summer flooding simply stimulated grass growth by compensating for any water deficit in the soil.

Today Turnastone Court is farmed by the charity the Countryside Restoration Trust. Although the irrigation system is long gone, the water meadows are still a flora haven. They remained unploughed even during the Second World War. Locals say that Mr Watkins stood at the gate of his main floodplain meadow and told the War Agricultural Executive that the field would be ploughed only over his dead body.

I have long been amused by the fact that the broken ditch that leaks over Lower Meadow makes a sort of poor man's irrigation system. The grass in that quarter of an acre is always greener. A sort of unintentional water meadow.

OCTOBER

Goldfinch

I LOOK FOR the changes in nature more closely in October than in any other month. Do many red haws on the hawthorn in the hedge really mean, as folklore says, there will be 'many snaws'? If the fieldfares arrive early will the winter be especially hard? Even though a profusion of berries indicates only the plant's past health I am hooked on weather divination. It is partly, I suspect, a primordial anxiety – shared with wildlife – that I need to prepare for the worst.

So of course the month begins with a diverting Indian summer, with morning sun-shafts in mist, and hoverfly (*Episyrphus balteatus*) on late-appearing buttercups. Starlings come up from the village with their party whistles to look for worms in the aftermath on the meadow.

I love everything about riding Zeb; the sea-deck motion, the creak of the saddle, the laughing excitement of cantering and galloping – and his pleasure in the same. I love the new perspective on old things one gets from the back of the horse. And above even this, I love that we are one, a unity; when American Indians first saw Spanish conquistadors on horses they believed them to be a single being.

The wild birds and animals of the meadow, for the most part, believe the same. We, the two-headed beast, amble around the perimeter and the fleet of rooks trawling the sea-green grass barely notices us.

It is different with the ewes, who stand rigid and watch me, then glance for a line of escape. When we near them on our walk, they lighten the load by squatting and peeing. Then run, bouncy-bottomed, to the far side.

Friar Tuck the ram strolls after them. Man-horse or horse-man he could not care, for he has a one-track mind. Fornication. October is the month of ovine sex in the country.

Friar Tuck stops and sniffs the ground where the ewes have pissed. Then he curls back his upper lip to show his teeth in a cartoon grimace. The flehmen response is not a male come-on but a means of closing his nostrils so he can suck air into the vomeronasal organ in the roof of his mouth. He is trying to detect the chemicals to know whether she is on heat. He himself is oozing so much testosterone that it hangs nidorous in the air.

One fat ewe with a ripped ear is clearly producing oestrogen by the pint. He licks her obscenely with a flicking tongue, then paws at her with a front leg, butts and bites her flank.

There are some try-out mountings.

She does not quite stand still; but then she does not run away either. He'll stay with her for the rest of the day, and cover her properly in the dark, the original one-night stand.

Tomorrow it will be a new girl.

Friar Tuck is not a wholly indiscriminate lecher. He likes his own Ryeland breed best. The Shetlands and Hebrideans will get covered last.

Rooks do not often visit the field, as they prefer the grain land at the bottom of the valley. Perhaps once or twice in autumn, boredom or a hungry memory of where bountiful worms are to be found brings them up here. There are twenty-three of them, garbed (or so it seems) in black cloaks. They feed into the brisk north wind, which blows over them, aerodynamically fixing them to the ground, as they stab it with their bone-white beaks. If they fed backside to the wind it would lift them up, and over.

The meadow is home; it is also a picnic site for visitors, and a stopping place for migrants on passage.

A place, too, for humans to reflect.

Humphry Repton in his *Observations on the Theory & Practice of Landscape Gardening* declared that 'the beauty of pleasure-ground, and the profit of a farm, are incompatible . . . I disclaim all idea of making that which is most beautiful also most profitable: a ploughed field and a field of grass are as distinct objects as a flower-garden and a potato ground.' Repton, along with Lancelot 'Capability' Brown, made the land into pictures instead of painting landscapes on canvas. Herefordshire was long a bastion of the gentry, and gentrified ideas about parkland percolated down to yeoman farmers. There's a Georgian farmhouse in Ewyas Harold with a ha-ha to the front, so the view over the rolling meadow is unbesmirched by a stock fence.

Tsar Alexander always considered that the next best thing to being the Tsar of all Russia was to be an English country gentleman. You can see why. They had the loveliest views in the world.

4 October A magpie sits on the Ryeland's back, pecking at its neck. This is biological symbiosis, ovine-corvine mutuality, despite appearances to the contrary; the magpie is picking ticks off the sheep's ears. The magpie gets a meal, the sheep gets cleaned.

The evenings are drawing in; in greyscale light I watch a black-eyed wood mouse lean up from a swaying hazel twig and pull a rose hip down, which it saws from its base with a flash of teeth. The rose hip tumbles down through the hedge to the ground, the mouse scrambling after it.

W

7 OCTOBER The last swallows on the telephone wires, chattering crotchets on a stave, the young ones gathering to await the nerve for the great voyage south. People used to think swallows hibernated in bubbles of air in ponds or down holes. In the early sixteenth century the Bishop of Uppsala, Olaus Magnus, in his *Historia de Gentibus Septentrionalibus* claimed that

> in the northern water, fishermen oftentimes by chance draw up in their nets an abundance of Swallows, hanging together like a conglomerated mass . . . In the beginning of autumn, they assemble together among the reeds; where, allowing themselves to sink into the water, they join bill to bill, wing to wing, and foot to foot.

His text was illustrated with a woodcut showing fishermen pulling the birds out of the water in their nets. Although Gilbert White proposed migration rather than hibernation (his brother, a chaplain in Gibraltar, saw swallows flying south over his head) he wondered about the very late broods, some of which were not sufficiently feathered to fly until mid-September: 'Are not these late hatchlings more in favour of hiding than migration?' White kept an open mind on torpidity, and hunted around the thatch of cottage roofs to find slumbering overwintering birds. To laugh at White is mere hubris; no one to this day is exactly sure where all the martin family overwinter.

The stuttery chattering and quivery flight of the swallow led medieval medicine to believe that, by association, eating the bird could cure epilepsy and stammering. A broth of swallow was the favoured medicine. The swallow was always a bird of goodness. Did not the medieval rhyme claim:

The robin and the wren
Are God Almighty's cock and hen.
The martin and the swallow
Are God Almighty's bird to hollow [hallow].

The swallows have gone; a chiffchaff passes through, stays for a day 'phoeeet'-ing and goes, the last summer migrant. The avian winter visitors have not arrived. We are now in the interval, when only native birds are here in the meadow.

This week of senile Indian summer heat underscores the sensuousness of the autumn world, its unaccustomed smells and fugitive scents. Gobstopper crab apples lie on the ground, rotting, vinegary.

10 OCTOBER Suddenly the weather hardens, a gale crashes down branches in the night, and I can feel the urgency of the grey squirrel in the copse as he or she violently scrambles about in the hazel bushes for nuts. I put on two jumpers in the morning, although something has apparently set fire to the trees; the coppiced hazel along the copse is burning gold from the bottom up.

W

12 OCTOBER The smell of woodsmoke from some distant fire. A blackbird gently 'spink-spink'-ing until it sees me, when it goes into full alarm-call mode; nothing says go away quite so fluently, so elegantly as a blackbird. There are now five blackbirds living in and around the meadow, three of them winterers from another place.

And there is frost already on the backs of the cows, at 5pm, their breath puffing white as they lie and chew the cud. Odd, and worrying, that the frost on the back of the venerable Margot, at twenty by far the oldest of our Red Poll, is thicker than on the hides of the rest of the herd.

And tawny owls in the woods and thickets along the misty stream are declaiming their autumn territories. A tawny owl never calls 'tu-twit-twoo'; the 'tu-twit' (actually, 'ker-wick') is the contact call; the 'twoo' (more accurately 'hoo-hoo-oooo') is the male's territorial call. If you hear 'ker-wickhoo', 'hoo-oooo', it is a duet, not a solo performance.

There are at least four owls calling in the declining of the day. September to November is when juvenile tawnies disperse, and tonight the young ones are trying to secure a fiefdom for food and for breeding. By wintertime pure they will either have succeeded or they will be dead.

The Anglo-Saxons knew the little goldfinch as *thisteltuige* or thistle-tweaker. They have a slightly sharper bill than other finches, made by Nature as a precision tool for extracting the seeds of thistles and teasels. The Latin for thistle is *carduus*, and informs the bird's scientific name, *Carduelis carduelis*.

The craze for keeping caged goldfinches came to a peak in the second half of the nineteenth century. In 1860, in Worthing alone, 132,000 birds were caught and the de-goldfinching of the countryside was an early concern of the Society for the Protection of Birds. Bedridden at the end of his life, John Keats also found himself in a sort of cage. One enjoyment in his life was watching goldfinches still at liberty, and he wrote in 'I Stood Tip-toe upon a Little Hill':

Sometimes goldfinches one by one will drop
From low-hung branches; little space they stop;
But sip, and twitter, and their feathers sleek;
Then off at once, as in a wanton freak:

Or perhaps, to show their black and golden wings
Pausing upon their yellow flutterings.

In autumn, the goldfinch, long since set free, joins in sociable flocks. There are at least thirty on the thistle band in the meadow, which I have grown (or allowed to grow) for them. The collective noun for goldfinches, a charm, is derived from the Old English *c'irm*, describing the birds' twittering song.

The autumn hawkbit, which is almost as yellow as a goldfinch's wing bar, has flowered in the promontory.

W

15 OCTOBER It might be cold but it is dry. The ground is holding up unseasonably well and I put the horses in the meadow for a change of grazing regimen. At night I go to check them. I see a shooting star, and the Milky Way bands the celestial dome. The high stars are limitless, and surely it is impossible that such a staggering show is not for my benefit.

And the stars come out tonight for me.

The horses grind their teeth while eating, and take extraordinarily long pisses. The sheep's eyes glisten green and jewel-like in torchlight.

Back in the house, I hunt out a verse from Thomas Traherne (1636–74), the Hereford-born metaphysical poet, that has tried unsuccessfully to escape from the vaults of memory:

The skies in their magnificence,
The lively, lovely air;
Oh how divine, how soft, how sweet, how fair!
The stars did entertain my sense,
And all the works of God, so bright and pure,
So rich and great did seem,
As if they ever must endure
In my esteem.

Traherne believed that man falls from a state of innocence because he turns from nature to a world of artificiality and invention. In *Centuries of Meditations* he advised: 'You never enjoy the world aright till the sea itself flows in your veins, till you are clothed with the heavens and crowned with the stars, till you so love the beauty of enjoying it you are earnest to persuade others to enjoy it too.'

18 OCTOBER Now comes the full cruelty of autumn. The haws loom scarlet in the hedge like the blood-drops of my fingers, where I have caught them in the thorns of the bramble. I have picked about two pounds of blackberries from the eastern side of the

hedge straddling Marsh Field, with the green bottle flies to guide me. Flies always sign the ripest fruits. Some of the on-the-turn blackberries are producing alcohol, and there are giddily drunk small tortoiseshell butterflies.

Bryony threads through the hedge in glorious orange and green chains. The sloes are full, and the rose hips overhang in sprawling, tempting arms. There is a feast of hedge fruit for wintering birds. But where are they?

Some familiar, overlooked friends are here beside me though. In flight, the pied wagtail utters a high-pitched two-note 'chiss-ick' sound; hence its joke-name of Chiswick fly-over for its habit of leaping past and calling as it goes.

Like yellow wagtails, pied wagtails feed predominantly on insects that they find while searching lawns, fields and verges. The insects are typically flies and caterpillars. On a sunny day such as this the dung from the livestock offers a smorgasbord. The bird's other names include molly washdish, nanny washtail and washerwoman, from its habit of feeding along the edge of ponds and streams. And, of course, because the tail bobs much like a laundress does when washing clothes. A finer or kinder eye describes their dainty femininity, and so they are *lady* dishwasher too. Their characteristic jerky gait rarely fails to bring a smile. As John Clare versed it:

little trotty wagtail, he went in the rain,
and tittering, tottering sideways he ne'er got
straight again.

Pied wagtails are a peculiarly British bird. Their close kin, the white wagtail of northern Europe, Russia and Alaska, has a definitely paler back.

W

21 OCTOBER Blowing an oceanic gale; it's hard to walk upright, my breath is panickingly taken away from me. The wind strips the leaves off the willow so they lie in belly-up shoals across the far end of the meadow. Acorns from the oaks are bombing the ground. These are toxic to cattle and horses. In the storm I lead the white-eyed horses out to the calm of the stables.

W

23 OCTOBER A marauding flock of wood pigeons roosts in the trees along the river at night, clappering out and back, as they settle. They feed on the acorns, including those of the twin oaks at the corner of the field. Although the flock is thirty or more strong it barely makes an impression on the shiny green mass of oak fruits lying on the field floor. I rake two barrow-loads up and feed them to the pigs.

In the wind jackdaws scatter and blow away.

As soon as I have left the field I see the jay fly in and pick up two acorns in its beak, then flit to the thicket, its progress signalled by its light-bulb rump. The jay is burying the acorns; it can bury hundreds a day as a precaution against inclemency. Some jays have been recorded as burying three thousand acorns and hazelnuts in a month. Probably half the oaks in Britain have been planted unintentionally by jays. The bird is a tree planter on a national scale. The bird's call is the sound of chalk being pulled down a blackboard.

W

26 OCTOBER While I seek shelter under the oaks from the rain, starlings fall like leaves on to the sodden earth, where the worms have been forced again to evacuate their burrows.

W

29 OCTOBER The first serious frost of autumn alchemises the field into an opaque white wasteland. There are perfect casts of ice in the old hoof marks of the cows. With numb fingers I pick mist-coated sloes for sloe gin.

NOVEMBER

Pheasant

THE RIVER IS so low I can step across its spindly width. A nuthatch scuttles down the elder, tapping for insects behind the Jew's ear mushrooms. At the back of Grove hedge, the crab apples have fallen into the ditch and gather slugs.

In medieval times it was believed that parent hedgehogs rolled on fruits and transported them home to their young. The hedgepiglets who lived under the pile of logs in the promontory will have no need of food. They are dead. Perfect scale miniatures of their parents, and grotesquely colourless in death. Only one has been eaten, scooped out of its spiny back; the other three seem to be unmarked. Presumably these died of cold. And a fox or badger is the perpetrator of the crime.

November is one of my favourite months, with its faded afternoons of cemetery eeriness, and its churchy smell of damp musting leaves. November suited perfectly poor crazed poet John Clare, who limned it so:

Sybil of months, and worshipper of winds!
I love thee, rude and boisterous as thou art;
And scraps of joy my wandering ever finds
'Mid thy uproarious madness.

Although some do feel, with Thomas Hood:

No warmth, no cheerfulness, no healthful ease,
No comfortable feel in any member –
No shade, no shine, no butterflies, no bees,
No fruits, no flowers, no leaves, no birds –
November!

Flint winds from the east cut us irrevocably from summer.

Hunger makes the hunter. From the house I see two male pheasants in the field, perambulating across with the dignity of Ming emperors. Even at hundreds of yards the low weak winter sun burnishes them into coppery magnificence. They are past the moult, they are in their feathered prime. Occasionally they stoop to peck at some flower or grass seed.

I get my shotgun from the gunsafe. By the time I reach the field, they have disappeared. I catch sight of their quick shadows in the copse but they slide away before I can get a clear shot. In the copse they are in their natural habitat, for what are pheasants but ornate jungle fowl? The first pheasants were brought here by the Romans, but probably did not go feral; *Phasianus colchicus torquatus*, the pheasant with the white ring collar, is an eleventh-century introduction. And after nine hundred years and annual releases of 30 million or so for shooting, the pheasant still looks gaudily out of place.

On a hunch I wait in the field, just under the overhang from the copse alder which bulges over the wire stock fence, loitering against it, the barrels of the gun nuzzled between neck and shoulder, more reassuring than a father's hand. With my left ear pressed hard to the trunk of the alder, I can hear every internal stress and strain as it shifts about.

The doleful day ticks down. The wren, cock-tailed capo of the copse, tells me off for loitering. When American poet Robert Lowell wrote 'For the Union Dead' and needed an image to explain the righteous Colonel Robert Gould Shaw, killed leading a regiment of black soldiers, he settled on 'an angry wrenlike vigilance'. I know precisely what Lowell meant; the wren continues its scolding staccato as I watch out over Lower Meadow while it prepares for bed. A tawny owl ker-wicks up the wooded stream, a robin warbles a few wistful bars from the Grove hedge, the sheep edge up the field to where it is highest, from where they have the best view of any approaching predators. The ancient Escley gurgles contentedly. The leaves of the hazel glow, the chill aches my face.

Then I hear the pheasants. A brief, proud 'cok-cok-cok'. They have left the copse and slinked next door to the Grove.

The light has gone to ashes. There are only seconds left in the day. I slip back the safety catch.

Up flies a pheasant from the Grove field, up, up, its tail streaming like the wake of a comet. I step forward and take the poacher's shot, the shot that is not for sport but for the kill. I shoot the silhouette just as it spreads its wings to break its speed before landing in the tree.

The bird falls thump into the field. Dead. As dead as though it had never lived.

The graffito blast of the shot is still echoing in the green valley, the blackbirds still squirting their alarm calls as I slip a length of baler twine around its neck to carry it home. The sheep, momentarily disturbed from their eating, put their heads to the field once more and carry on mowing.

The smell of gunpowder is thick around me, and masks out even the rotting incense of the autumn leaves. A child's full moon is struggling to break through the gloom.

How did I know the pheasant would roost there, on that bare branch in the alder? It is where I would have chosen to sleep if I were a pheasant, a place too high for foxes but not so dense with leaves that I could not see into it.

Rationally it seems fair, even appropriate, that if one farms for wildlife one can eat the wildlife. The justification does not stop me suffering the agonies of sentient killers and those unforgiving lines from Blake's 'Auguries of Innocence' start to spool:

To see a World in a Grain of Sand
And a Heaven in a Wild Flower,
Hold Infinity in the palm of your hand
And Eternity in an hour.

A Robin Redbreast in a Cage
Puts all Heaven in a Rage.
A Dove house fill'd with Doves & Pigeons
Shudders Hell thro' all its regions.
A Dog starv'd at his Master's Gate
Predicts the ruin of the State.
A Horse misus'd upon the Road
Calls to Heaven for Human blood.
Each outcry of the hunted Hare
A fibre from the Brain does tear.
A Skylark wounded in the wing,
A Cherubim does cease to sing.

The Game Cock clipt & arm'd for fight
Does the Rising Sun affright.

And on, on until:

Kill not the Moth nor Butterfly,
For the Last Judgment draweth nigh.

W

6 NOVEMBER A squirrel walks towards me through the grass of the meadow. The dogs bark on the yard at the arriving postman, which alerts the squirrel; then he sees me and disappears in a mercury rush. A flight of ducks (their dart-shape issues a squeak – mandarins) comes up off the river. I am checking the rackety troublesome riverside stock fence, to stop the ram escaping. The badgers have pushed under the fence to eat the acorns.

W

The cattle are back in the field, taking a last 'bite' of the grass growth brought on by a sequence of muggy days. Then it rains, and quickly the ground becomes too soft to support their weight; in the jargon of farming they will 'poach' the ground, turn it into a sea of mud. In a squinting downpour I herd them to their winter quarters.

But the remains of the autumn sun set November alight again. The days brighten up from elements of sun snagged in the spiders' webs in the hedges and tussocks. The ash has been denuded by the downpours, but the alder and oaks are still holding on to their greenery.

W

I like the Braille of bark, the way that – with eyes closed – one can identify a tree by touch. Oak trunks have rectangular mosaic tiles; the old ash has latticework for skin; the slumped elder by the brook gnarled longitudinal fissures in polystyrene; silver birch the smoothness of silk stockings. Then there is the last hazel in the Bank boundary, gone from shrub to tree in the hedge that is no longer a hedge but a plodding parade of bowed single sentinels; it is worn smooth and polished by cows' rubbing over ancient summers. They have done the same to the two oak-trunk gateposts, as they have barged and passed by, so that the bleached-out wooden pillars are glossy to sight and feel.

Margot will no longer be adding her polishing. The great beast is dead. I suppose, as this is a death announcement, she should be titled properly: 'Worlingworth Margot, daughter of Woldsman King Harry and Woldsman Ember'. She was a Red Poll cow with pedigree. She has been arthritic for two years, and has fallen over from time to time, always to be hauled upright by many hands or the jeep, to plod on happily behind the rest of the herd.

This morning there is no Lazarus moment, there is no miracle. She has fallen into the paddock ditch, and it takes the tractor – fumes powering out of the bonnet exhaust – and a chain, the industrial one with links the size of fists, to get her out of the sucking mud. But hauled and beached in the field, even my words of love will not make her rise. She is lying on her side: her eye, white and marble and veiny, stares up. She is dressed embarrassingly, a grim mud shawl over her gorgeous coat. Her struggling front hooves cut small crescents in the sward, so there is no grass, there is only more mud.

Her daughter, Mirabelle, moseys over, and noses at her mother. She can smell death on the mist.

I begin to walk to the house to call the vet to administer death by injection, then stop. Margot hates vets, with their personal deodorant of ointments and ailments. She is an old lady expiring naturally, and I let her go from the world this way. All the wan morning long her daughter stands next to her, but never again looks at her. The other cows one by one pay their curious, sniffing respects. I am there at the last, when shit and life leave her. She dies in an afternoon sky painted in heavy purple oils. I cover her face with a plastic sugar-beet sack so the crows won't peck her eyes out.

Margot. My lovely, cantankerous old cow, a true beast of the field.

All cattle are descended from as few as eighty animals that were domesticated from wild ox or aurochs in Iran some 10,500 years ago, according to recent genetic studies. This is not long after the invention of farming. The history of farmed cattle in England is more recent, with the first domestic cattle arriving here about six thousand years ago. They made an immediate impact on the earth of Albion; the dung beetle population increased exponentially.

Sheep were the main domestic grazer, especially outside fertile lowland valleys, providing wool, meat and milk. Cattle, however, had one great advantage over sheep; they could be used for locomotion. There are cattle bones in Neolithic sites showing the stress-induced damage that comes from hauling and ploughing. A cow, of course, is also an 'ox'. Poor nutrition or selective breeding made domestic cattle smaller (some in Scotland were only a metre high at the shoulder), and easier to handle than their wild counterparts.

The laws of the Saxon King Ine of Wessex

(688–95) show cattle farming to have reached a stage not unlike today, with the beasts contained within fields.

> 40. The landed property of a ceorl shall be fenced both winter and summer. If it is not, and if his neighbour's cattle come through an opening that he has left, he shall have no claim to such cattle, he must drive them out and suffer the damage . . .
> 42. If free peasants have the task of fencing a common meadow or other land that is divided into strips, and if some have built their portions of the fence while others have not, and if their common acres or grasslands are eaten [by straying animals], then those responsible for the opening must go and pay compensation to the others, who have done their share of the fencing, for any damage that may have been suffered.

Breeding cattle for beef was a later but wholly English invention, one in which the meadows of Herefordshire played a leading role. Until the eighteenth century, the cattle of southern England were wholly red with a white switch, similar to the modern Red Poll. During the eighteenth century other cattle (mainly Shorthorns) were used to create a new type of beef cattle with a characteristic white face, the Hereford, which from 1817 sold to the world, from America to Australia. Nothing – not St George, rugby, cucumber sandwiches, cricket on the green – is as English as beef. It has been a national symbol for centuries, and for the French we are *les rosbifs*. The song 'The Roast Beef of Old England', penned in 1731, was once a national anthem, sung by the audience in theatres.

When mighty Roast Beef was the Englishman's food,
It ennobled our brains and enriched our blood.
Our soldiers were brave and our courtiers were good
Oh! the Roast Beef of old England,
And old English Roast Beef!

By far the most important factor in the taste of beef is the animal's diet. Unlike in America, English cattle are still largely fed on what they would naturally eat – grass and fodder made from it, like silage and hay. Grass-fed meat always proves tastier in comparison tests and it's certainly healthier, with more vitamins, less unhealthy fat and beneficial Omega-3 fatty acids.

But it doesn't taste as good as it did. How could it? Animals are what they eat. The varied, rich herbage of yore gave flavours.

The very last aurochs died in Poland in 1627.

(Mankind's destruction of other species by hunting is nothing new.) An auroch was the size of one of the big cows of today, like the Belgian Blue. So, after ten thousand years we have only succeeded in producing cows as big as the wild ones.

W

Walking disconsolately around the meadow the evidence of Margot is all around. (I pay attention to livestock shit the way Roman augurs inspected owl innards. An awful lot about the state of an animal can be determined by its excrement. Inky-black 'crotties' that break down into pellets are a good sign in sheep; green goo is bad, and probably betrays too high a burden of intestinal worms.) Where there's muck there's a wealth of invertebrate life. Cow dung contains up to 250 insect species. Of the 56 British Red Data Book species of beetle associated with dung, 16 live in cattle excrement; 15 in horse, 13 in sheep. The excrement from one cow can produce food for 0.1 ton of insect larvae per annum.

Dung is such a popular food source that insects may be laying their eggs in it before it has even hit the ground. Grazing only extracts about 10 per cent of the energy from grass, leaving the comminuted waste enriched. All of these invertebrates help to break down and recycle the dung, as well as providing a bonanza of food for other predatory animals. Dung manures the earth, and feeds a whole long chain of life.

The oddity is that some of the most beautiful of all insects live in this dung. The afternoon is hot enough for sultana-brown dungflies to loiter. I poke about with a stick in some of the cow pats. There are rove beetles and dor beetles, including *Geotrupes stercorarius*, black on top with a fetching purple rain-sheen underneath.

Unseen and unhonoured labour the bacteria, about a billion of them per gram, the land's hidden farmers, breaking down the faecal matter into humus, into soil.

11 NOVEMBER Martinmas. The day the livestock was slaughtered and salted. Remembrance Day, the day when the French light the darkness with lanterns. The Feast of St Martin, Martinmas was a time for celebrations with great feasts and hiring fairs, at which farm labourers would seek new posts:

> If they seek a new place the men and boys stand in the street of the town, often with a straw in the mouth, but when engaged they join the merry throng amongst the booths and shows of the fairground. The women and girls have

generally a hall provided for them by the ladies of the district, and suitable refreshments are supplied. In the evening there is a dance and often other entertainments. This year in most centres the difficulty was to get a servant of any value, as masters were so fearful of losing their capable men or women, that they hired them on. Under these circumstances second-class servants got a better wage than they had expected. In the average of cases, taking a number of the chief fairs, head men at £19 for the half year got an advance of a pound on the wage of a year ago. Experienced dairy-maids at £15 also secured an advance.

Manchester Guardian, 22 November 1913

The traditional food eaten on Martinmas was beef. Martlemass beef, dried in the chimney, was the staple winter diet in the grass parts of Herefordshire. It was claimed:

Marlemass beef do bear goode tack
When country folk do dainties lack.

Since 1918 the 11th has been commemorated as Armistice Day, and all remnants of the old Martinmas celebrations have disappeared.

On this Remembrance Day Margot is taken away, her stiffened body winched into the back of a cavernous lorry by the slaughterman, to play her role in a vision of Hell by Goya on downers, to lie among the bloated sheep, wood-legged cows, a yellow pig.

I would like to bury her where she belongs, in the field, so her flesh nurtures the flesh that is soil. Instead, by government order, she has to be incinerated in an abattoir.

I could weep.

18 NOVEMBER The fieldfares and redwings have been drifting through for weeks, but now they arrive in Viking hordes. They are the sound of winter.

I heard them come in the night, 'chakka-chakk'-ing. Sure enough, in the morning they are in the orchard scrumping through the windfalls, fifty or more of them. Down in the field, there is another party of the Norse thrushes, all mixed together, stripping the haws by the gate.

The fieldfare takes its name from the Anglo-Saxon *felde-fare*, 'the traveller over the fields'. For Chaucer, *Turdus pilaris* was the 'frosty feldefare' and harsh weather does indeed drive them south from the Scandinavian homeland. A million redwings and fieldfares come down from the north. Most of them seem to have arrived with us. Tippi Hedren would not like it.

W

20 NOVEMBER Note on paper: 'Alder has lost leaves but bees [catkins] still there. Oak leaves [look] burnt, the younger oak in the Marsh Field a funeral sail in wind.'

W

21 NOVEMBER Picking rose hips from the straggling dog rose by the thicket I come across one of those extraordinary abnormal growths that is a Robin's pincushion, which is caused by the tiny gall wasp called *Diplolepis rosae*. The female wasp lays her eggs in the bud of the rose, and in so doing re-codes the normal process of flower growth to produce a round moss-like ball. Inside is a honeycomb of chambers, in each of which is the grub of the wasp. Also in residence is a motley community of opportunists, including other gall wasps, plus parasitic wasps. There is even a species of chalcid wasp that parasitizes parasitic wasps that parasitize parasitic wasps – in total, a chain of four parasitic wasps feeding on each other.

This pincushion is past its prime; yet broken open in the depth of November it is still housing minute wasp maggots.

Some long-tailed tits are busy in the oak. There

must be about twenty of them, and they may well all be related. They constantly call to each other, 'tsi, tsi'. They are a bird whose habit of cooperation would shock Darwin.

24 NOVEMBER A dead badger on the lane; I am certain it is the old boar. The victim of a car. Badgers are not cute to look at: the pig snout and pied facial stripes are weirder still up close.

I drive on, leaving the badger on the verge of the lane. There is nothing as lonely as death.

The next morning I go back with a plastic feed sack (from the inevitable beet pellets) to collect the badger and bury him in the field.

The body has already been removed, probably by the Council, who have someone who collects badger corpses. But I like to think, in a wave of sentiment, that the badger's family took his body and buried it. The naturalist Brian Vesey-Fitzgerald once saw a badger funeral. It was 1941, the middle of World War Two, and the badger family dug a grave, dragged and shoved the deceased into it, then covered it with earth. Dust to dust. The sow wailed throughout the moonless night.

27 NOVEMBER Tonight I see something I have never seen before, something I never even knew of. It's late, and I have gone for a moon-time walk around the fields, because I love the solitude of the dark. While I am looking to the west and the unbroken night of mid-Wales, an arch of white light suddenly appears in the sky and spans the earth before me. I feel afraid, as though I have been singled out for some almighty moment of revelation, that I have been entrusted with some Damascene vision, and several seconds pass before I understand what it is I am looking at.

I am looking at a rainbow at night. A moonbow.

28 NOVEMBER Note: 'Morning in the field: jackdaws at 6.45am in great squadrons, wheeling. More join in. Lots of noise. Then fly off en masse, aside from a few persistent individualists who go their own way.' Two hours later there is mother-of-pearl sunshine.

There are birds calling greetings and alarms, but in the field only the robin *sings*; even in winter the robin will defend its smallholding. For all its charm as the pin-up of the Christmas card industry, the robin is a vicious little thrush. A dead robin lies on the grass beside the thicket, its head battered from pecking, one eye burst by a torturing, puncturing beak.

W

This is a dying world. A nearby farm is diversifying into holiday accommodation. Their field of the beautiful aspect will grow tipis. Which is like a dog shitting on a white Berber carpet.

The wind rakes the valley, searching into every fold of earth and unbuttoned flap of coat. There is Reynard the dog fox digging excitedly. His fur is in bloom and has the scorching-red hue of a fire ember. But so strong is the wind that it is ruffling his hair; a cartoon fox plugged into the mains would look sleeker. He does not hear or see me as I walk up behind him. The soft ground absorbs the vibrations from my feet. Childishly I cannot resist giving him a fright, and when I am almost within touching distance of his white-tipped brush I cough loudly.

Foxes can run.

W

30 NOVEMBER Warm, orange glow in the afternoon. The sigh of my feet in the frosted night grass. Wrap my coat closer, wrap myself into the ground, fold myself into the earth. As night descends I can hear the shiftless hunting of voles, shrews and mice in the hedge. Shrews do not hibernate, as they are too small to store

fat reserves sufficient to see them through the winter. And spangled is the only word for this starry night of seeping cold.

DECEMBER

Fox

THE FIELD IS DEAD. A single snap. A soundless photograph. An inverse of the bustle of summer. The grass has stopped growing.

A black-and-white picture too with this lifeless mist. In the bottom of the Grove ditch the red campion that clung tenaciously throughout the wrong season has, at last, given up. Only in the thicket and hedges do I find colour, in the red of the holly, hawthorn and rose hips. In medieval times the holly was the Christmas tree, its scarlet berries held to be the resemblance of Christ's blood. There are other reds in December. Robin. Hunting jacket. Fox.

The mist wanders away, to be replaced by an exhilarating week of hoar frosts and blue skies, and the barking of a fox at the crescent, huntsman's moon. Venus is out in the sky even before the sun goes down over the mountain. Then it snows at night.

In the morning, there is the peow of the buzzard over the bed-white field and a tell-tale streak of urine on a tussock. Through a gap in Bank hedge I can see the fox walking gently over the snow, ears alert. The fox stops, turns its head, all the better to hear with. Takes another few paces forward. Listens. Then rears up on its hind legs and does a diving pounce.

A flurry of digging snow.

A caught field vole, which is swallowed whole.

When I was about ten my father brought a redhead home in the back of his yellow Rover 2000. The redhead, on closer inspection, turned out to be the stuffed fox that graced the window of the gunsmith's in West Street in Hereford. The fox was more than stuffed, it was the centrepiece of an Edwardian field sports monument, a complete scene in which the snarling fox looked down on two rabbits emerging from a burrow. (The edifice was constructed from some sort of early painted plastic.) The gunsmith's shop was closing down, and my father, sweetly, had thought I would like the fox.

I did. So did our Labradors when it was installed in my bedroom; they used to sneak in and chew the stuffed rabbits.

I would spend hours gazing at the fox. I would measure it, from clee to shoulder, the tip of canine tooth to jawbone, from the end of the brush to the back. The fox set me off on a trail of nature reading from the public library in Broad Street, during which (with the guidance of a master from school) I stumbled upon *Wild Lone: The Story of a Pytchley Fox* by BB. In a sense I owe that stuffed fox a great deal, because it was BB who, above everyone, inspired me with a spiritual respect for nature, as opposed to simple admiration or sentimental regard. (Though I can do both of those too.)

While I liked my stuffed fox, I have never been quite so keen on the real thing, the killer of our chickens, ducks and lambs. Respect but not love. My ambivalence is perfectly incorporated in the fact that I am the only person I know who has both hunted foxes on horseback and 'sabbed' fox hunts.

Only today do I fully understand why foxes make me uneasy. For a canid, the fox is disconcertingly catty. Aside from the feline mouser pounce, foxes have vertically slit pupils. The Edwardian taxidermist who stuffed my fox gave it appealing amber dog eyes.

The fox trots off through the snow, aware of his handsomeness. He is a proper country fox. Given half-decent light I can distinguish him from all others by his size, and the possession of the purest black legs. He is three, and the father to the cubs born back in February. Sometimes I watch him from the house when he is on a circuit of his territory, which is, roughly, two and a bit farms, about a hundred acres, the main borders being the river and the road. Although he trots, his progress is slow since every fifty yards or less he stops to scent. Reynard, as I always think of him, is doing well to reach three. The mortality rate for adult foxes is 50 per cent per year. The mortality rate for cubs and juveniles is 60–70 per cent.

I know that one of the cubs from the copse is dead; it crossed the river and entered the territory of the quarry wood foxes. The body was in plain view in the middle of the sheep field across from the finger two days ago. When I reached it, I could see that it had been badly mauled around the neck.

Of course, my contradictory feelings towards foxes are a national tic. No other animal has been so strenuously exterminated or so earnestly anthropomorphized – and frequently as Reynard, the name coming originally from a twelfth-century Latin poem in which Reinardus torments his dim lupine uncle, Ysengrimus. He appears in Geoffrey Chaucer's 1390 *Nun's Priest's Tale* as Rossel and fully fledged as Reynard in William Caxton's *History of Reynard the Fox* published in 1481. Systematic hunting of the ingenious loner, *Vulpes vulpes*, began similarly early; Edward I had royal fox hunters during the thirteenth century. Killing of foxes was not restricted to hunting with dogs; under the Tudor Vermin Acts the head of a fox had a bounty of one shilling, and no one was fussy how Reynard was killed. According to the animal historian Roger Lovegrove in his book *Silent Fields*, the ancient countryside of Britain, including these Welsh Marches, is where fox hunting was most energetically pursued.

All the green, billiard-table smoothness of the meadow is gone; two days of hoar frost have left the grass pallid and listless. A moment of beauty is given by raindrops on the sedge heads glistening like rubies and emeralds in a Rider Haggard story. The hedges and trees in the copse are skeletons of their former selves. The buzzards greet the dawn, the jackdaws close the day.

14 DECEMBER My daughter's school carol service, in Hereford Cathedral: the first part of the service is held in the dark, save for a ring of isolated candles hanging above the middle of the transept. Later that night, I go down to the field, and stand there in the vertiginous dark, with the lights of the stars above me. The mountains make for high walls, the stars for candles. There is no difference between the cathedral and the field.

I rage against the coming of artificial light. To stand in an immense starred night is to be a citizen of the universe. To see its immensity. Stars made Australopithecus gaze up in wonder and dream. London has not seen the stars since the Blitz.

The next afternoon, a cheery troupe of starlings

travels over from the village. They are bright with their winter plumage, which itself is a stars-in-the-night design.

16 DECEMBER The field underground: I'm digging out a hole for a fence post after a drizzle that has helpfully defrosted the land. I find digging into the meadow intoxicating, the way it reveals the meadow's hidden underself.

There is a surprising amount of life in the bottom of a tussock: a C-shaped cockchafer grub, a wolf spider, minute club-headed yellow fungi, the pupae of butterflies, then as the spade goes in you can see all the workings of the worms, each an organic plough pulling leaf matter down, composting, constantly sending casts to the surface. The workings go down 40cm or so, deeper even than the swollen roots of the dandelion. Sap rises in the spring; the goodness of plants sinks into the roots in winter.

A heavy Sunday afternoon with an unmoving off-grey sky, a chainsaw gnawing away far upstream. Edith and I take a turn around the field, around the edge. As we near the newt ditch Edith puts up a snipe, which zigzags off.

Back in the house, I look through my diary and write down all the birds I have seen in the field this year. First, those on the ground or in the hedges and trees:

snipe, song thrush, blackbird, chaffinch, robin, buzzard, bullfinch, raven, magpie, skylark, curlew, wood pigeon, goldfinch, rook, pied wagtail, common partridge, nuthatch, spotted flycatcher, chiffchaff, blackcap, greater spotted woodpecker, green woodpecker, wren, long-tailed tit, redwing, fieldfare, meadow pipit, house sparrow, jackdaw, mallard, yellow wagtail, carrion crow, blue tit, great tit, lapwing, heron, house sparrow, tawny owl, barn owl, little owl, yellowhammer, starlings.

Flying above or alongside: red kite, kestrel, mandarin duck, heron, canada goose, swift, swallow, house martin, merganser, kingfisher, sparrowhawk, great crested grebe, dipper.

There are two more species overall than last year, though there are a couple of conspicuous absentees: tree sparrow and brambling.

In the mood for lists, I also collate all the flowers in the meadow: cuckoo flower, pignut, yellow rattle, meadow vetchling, tormentil, bird's-foot trefoil, bugle, meadow saxifrage, devil's bit scabious, eyebright, white daisies, red clover, white clover, bluebells, lords and ladies, red campion, yarrow, dandelions, yellow archangel, foxgloves, camomile, thistles, stick mouse-ear, ragged robin, ground ivy, cleavers, wood anemone, dog's mercury, stitchwort, dock, cow parsley, hogweed, cowslip, primrose, wild roses, honeysuckle, bush vetch, ribwort plantain, hairy bittercress, rough hawkbit, common meadow rue, nettles, meadow cranesbill, autumn hawkbit.

17 DECEMBER Through the field blows a relentless wind. Now the foliage has died back or been torn away, smaller birds are more visible. In the lichen-encrusted apple trees in Bank Field a tiny treecreeper ascends, pecking with its beak, which is decurved like the curlew's.

A fast-darkening day, and a badger appears by the copse. Almost instantly three travelling carrion crows mob it, and it disappears. When I go over, I see that the badger has dragged some uncollected hay towards the fence, presumably for bedding. This is ambitious, since the sett is hundreds of yards away. But clearly life goes on in the sett, as it has done for countless years past.

W

19 DECEMBER One of those curiously mild, almost spring days that December can throw up. In the morning the entire field is woven with spider webs, so much so that the low sun reflecting off them is blinding, like moonlight lying on the sea. The air fills with drifting gossamer strands, each carrying a juvenile spider leaving home.

W

23 DECEMBER It is getting towards prime mating time for the foxes, and the night is alive with barking. Foxes have a wide range of vocalizations, though they mainly 'yip' in a staccato style or bark 'woo-woo'. They also emit a 'waaaaaaaaaaaaaa' howl to raise the hairs on the neck.

I walk down to the field in the moonlight. Now I can hear another fox sound: 'gekkering'. This is a clicky chattering interspersed with squeaks, more parrot than dog. It is used in aggressive encounters. The noise is coming from the river, and is discernible even above the background babble. My advance across the field is hardly silent: two days of rain have left the clay sodden and this has now frozen. The moon's rays pick out flickering diamonds in the grass. Peering over the bankside fence I can see the silhouettes of two foxes either side of the river, eight feet apart. Red foxes are highly territorial. I skirt out into the promontory and around. They are so preoccupied, and my loud advance sufficiently confused by water noise, that I get to within ten feet, and the fox on the shingle across the Escley is perfectly illuminated; I can see the breath from his snarling mouth, the flatness of his ears.

Foxes have been doing this for a long time. Remains of the red fox have been found in Wolstonian glacial sediments from Warwickshire, meaning they were around between 330,000 and 135,000 years ago.

Something in the atmosphere changes, and the fox on the far bank looks up and sees me, and bounds off. Reynard disappears into the shadow of the thicket.

Then I go to my earth.

27 DECEMBER At night the temperature dips below freezing. The pulse of life is stilling, slowing. Over the field sits the rusty-iron smell of the year's finish. The field would be an almost empty scene if I could not see my memories superimposed upon it. This is the field I cut by hand, the field I was part of, where I learned the pleasure of simple things. If you want to know what happiness is, ask the fellow who cut the hay.

31 DECEMBER New Year's Eve. I put the sheep in the field, and a load of hay in their feeder. In a sort of virtuous circle the hay is from this field – the hay I made in the summer, pulled down on those now indispensable tarpaulins.

The sheep and hay are in. The raven croaks. I shut the gate and leave. This is how it is, has been, how it shall be evermore.

Flora

alder
apple
ash
autumn hawkbit
bird's-foot trefoil
blackberry
blackthorn
bluebell
bracken
bramble
bryony
bugle
burdock
bush vetch
camomile
cleavers (goosegrass)
cock's foot
common bent
common meadow rue
common vetch
cow parsley
cowslip
crab apple
crested dog's tail
cuckoo pint
dandelion
deadly nightshade
devil's bit scabious
dock
dog rose
dog violet
dog's mercury
dyer's greenweed
elder
elm
eyebright
field forget-me-not
field maple
field scabious
fir
foxglove
goat willow

ground elder
ground ivy
hawthorn
hazel
hemlock
hogweed
holly
honeysuckle
ivy
Jack-by-the-hedge
Jew's ear mushroom
knapweed
lady's smock
lesser celandine
liberty cap mushroom
lords and ladies
marsh thistle
meadow buttercup
meadow cranesbill
meadow fescue
meadow foxtail
meadow grass
meadowsweet
meadow vetchling
mistletoe
mouse-ear
nettle
oak
pignut
primrose
quaking grass
ragged robin
red campion
red clover
red fescue
ribwort plantain
rosebay willow herb
rough meadow grass
rye grass
saxifrage
sedge
sloe
snowdrop
sorrel
St George's mushroom
stitchwort
sweet vernal
thistle
timothy
tormentil
tufted hair-grass
turf mottlegill mushroom
velvet shank mushroom
white clover
white daisy
willow
wood anemone
woodrush
yarrow
yellow archangel
yellow rattle
yellow waxcap mushroom

Fauna

aphid
backswimmer
badger
barn owl
bilberry tortrix moth
black slug
blackbird
blackcap
blackfly
blue tit
brimstone butterfly
brown-spot pinion moth
bullfinch
bullhead
bumblebee
buzzard
cabbage white
caddis fly
Canada goose
carrion crow
chaffinch
chalcid wasp
chalk hill blue butterfly
chameleon frog
chiffchaff
cockchafer
common blue butterfly
common partridge
crane fly/leatherjacket
cuckoo
curlew
damselfly
Daubenton's bat
dipper
dor beetle
dragonfly
dungfly
dunnock (hedge sparrow)
earthworm
field mouse
field vole
fieldfare

fox
fox moth
frog
frog-hopper
gall midge
gall wasp
gatekeeper butterfly
glaucous shears moth
goldfinch
grass snake
great crested grebe
great spotted woodpecker
great tit
greater horseshoe bat
green bottle fly
green woodpecker
Hebrew character moth
hedgehog
heron
horsefly
house martin
house sparrow
hoverfly
jackdaw
jay
kestrel
kingfisher
ladybird
lapwing
large-jawed spider
leaf bug
least yellow underwing
lesser cream wave moth
little owl
loach
long-tailed tit
magpie
mallard
mandarin duck
marsh fritillary butterfly
mayfly
meadow ant
meadow brown butterfly
meadow grasshopper
meadow pipit
meadowsweet
merganser
merlin
midge
minnow
mole
money spider (5)
mosquito
newt
nightjar
noctule bat
nuthatch
orange-tip butterfly
otter
palmate newt

peacock butterfly
pheasant
pied wagtail
polecat
pond skater
powdered quaker moth
rabbit
rat
raven
red grouse
red kite
redwing
robin
rook
rove beetle
satyr pug moth
scarce vapourer moth
short-tailed vole
shrew
silk cell spider
six-spot burnet moth
skylark
slug
small copper butterfly
snipe
soldier beetle
sparrowhawk
spotted flycatcher
springtail
squirrel
starling
stoat
swallow
swift
tawny owl
thrush
toad
tortoiseshell butterfly
treecreeper
trout
violet ground beetle
water boatman
water vole
weasel
willow warbler
winged agate (flying ant)
winged meadow ant
wolf spider (2)
wood mouse
wood pigeon
woodlouse
wren
yellow wagtail
yellowhammer

A Meadowland Library of Books and Music: a List Raisonné

You are what you read. So I offer the following in explanation.

I wish I could remember which master at school read *The Little Grey Men* to us (stories about gnomes for twelve-year-olds, absurd!), though I think it must have been the keen, young Mr David, who wore the bottle-green cord jacket. While BB's gnomes amused me, they did not interest me as much as the book's subtext: the natural history of the British countryside.

I was not a stranger to that countryside; it was outside the door of my house, it was what my grandparents farmed, it was what my family had lived in (the Herefordshire variant) for around nine hundred years. What BB did was make me see interconnectedness. If gnomes could communicate with wild animals, then why not me? It was a way of thinking about nature which is not Us and Them, but We together. It was the natural world from the inside out, not from the outside in.

I was already a 'nature boy', given to rambles with my black Labrador dog, Rover (of course), usually alone, though sometimes with cousins or friends. My standard equipment, aside from the dog, was an outsize pair of Boots Empire 10 x 50 binoculars, which always swung hurtingly when I climbed trees to peer into birds' nests, especially the twig shanties of wood pigeons in the tops of the elms. I still have my *Observer's Book of British Birds' Eggs*, given to me at seven, just as I have the *Observer's Book of British Birds*, of *Wild Animals*, of *Pond Life*. Plus the *AA Book of Birds* which was my tenth birthday present from Auntie Eileen and Uncle George, which sits on the bookshelves next to those other indispensable bird-identification guides from the 1970s: *The Hamlyn Guide to the Birds of Britain and Europe* and the *Collins Pocket Guide to British Birds*. The latter was a school prize, won in a surprising and never equalled moment of scholarly success.

No sooner was I introduced to BB than I was off to Hereford Library, to come home with my school briefcase fat with *Down the Bright Stream*, *The Wild Lone* and *Brendon Chase*. It is the last two of these books, really, which means that BB is the author to have most influenced me. *The Wild Lone* is still, to my money, the best attempt to get inside an animal's head and life, while *Brendon Chase* is the classic of boy's adventuring in the British countryside, the story of the Hensman boys living wild in the woods. (I admit Arthur Ransome's

Great Northern? runs it close.) As a very big boy in my forties I finally got my chance to live wild, spending a year surviving on what I could forage and shoot, an experience recounted in *The Wild Life*. Being outside should never be anorak-dull. Being a bird watcher should not mean one is predestined to be a character in a Mike Leigh play. The outside should always be an ecstatic experience.

If BB (properly Denys Watkins-Pitchford) remains the nature writer I most admire, others have stuck to me over the passage of time, rather like goose grass sticks to the coats of sheep as they pass along hedges. The inherent anarchist in me loves William Cobbett's self-sufficiency classic *Cottage Economy*, the middle-aged conservative shooter adores Brian Vesey-Fitzgerald's *British Game*. (The latter book was Number 2 in the New Naturalist series, any and all of which deserve a place on a nature lover's bookshelf.) I have made both my children read Barry Hines's *A Kestrel for a Knave* – I read it too as a child – to make them understand the privilege of having a pet. You are never lonely if you have the love of an animal.

George Orwell once wrote to the effect that autobiography should never be trusted unless it reveals something outrageous. My small confession is that I distrust science, and in my science-free A levels I discovered English pastoral poetry: Thomas Hardy, John Clare, Edward Thomas and the honorary Briton, Robert Frost.

(Thomas and Frost had lived as part of the circle of Dymock Poets, only miles from where I grew up.) Don't they all communicate truths about nature too?

You read what you are. I have the bad habit of reading books about farming from another world: Britain before DDT. George Ewart Evans's *Ask the Fellows Who Cut the Hay*, John Stewart Collis's *The Worm Forgives the Plough*, George Henderson's *The Farming Ladder* are always reassuringly close.

The full list of meadowland books to be found on my own shelf is:

Richard Adams, *Watership Down*, 1972: the lapin *Aeneid*.

J. A. Baker, *The Peregrine*, 1966: Baker's account of a year spent following peregrine falcons, in which he elides the distinctions between man and bird, won the Duff Cooper prize for 1967.

BB (Denys Watkins-Pitchford), *The Wild Lone*, 1938; *Manka the Sky Gypsy*, 1939; *The Little Grey Men*, 1942; *Brendon Chase*, 1944; *Down the Bright Stream*, 1948

Ronald Blythe, *Akenfield*, 1969: the last days of traditional agriculture in Suffolk.

Maurice Burton, *The Observer's Book of Wild Animals*, 1971

Geoffrey Chaucer, *Parlement of Foules* (trans. C. M. Drennan), 1914

A. R. Clapham, *The Oxford Book of Trees*, 1986
John Clare, *The Shepherd's Calendar*, 1827: the countryman's year, tasks, festivals, rhythm, in verse.
John Clegg, *The Observer's Book of Pond Life*, 1967
William Cobbett, *Cottage Economy*, 1822: the original classic of self-sufficiency. *Rural Rides*, 1830: a splendidly splenetic and entirely accurate portrait of Georgian England from the back of a horse.
John Stuart Collis, *The Worm Forgives the Plough*, 1973: working in the British countryside during WW2.
Country Gentlemen's Association, The Country Gentlemen's Estate Book, 1923
R. S. R. Fitter, Collins *Pocket Guide to British Birds*, 1973
Roger Deakin, *Wildwood*, 2007
G. Evans, *The Observer's Book of Birds' Eggs*, 1967
George Ewart Evans, *Ask the Fellows Who Cut the Hay*, 1956: the oral history of a Suffolk village.
Thomas Firbank, *I Bought a Mountain*, 1959: hill-farming in Wales.
W. M.W. Fowler, *Countryman's Cooking*, 2006: originally published in 1965, and quite fantastically politically incorrect.
Sir Edward Grey (Grey of Fallodon), *The Charm of Birds*, 1927: the man who led us into WW1 was an ardent ornithologist; this is his study of birdsong. His eyes were in decline, his ears were perfect.

Geoffrey Grigson, *The Englishman's Flora*, 1955: always the *Flora* I turn to.

Lt-Col Peter Hawker, *Instructions to Young Sportsmen*, 1910: possibly the most popular book ever on shooting.

George Henderson, *The Farming Ladder*, 1944

Otto Herman and J. A. Owen, *Birds Useful & Birds Harmful*, 1909: This was the first book on birds I ever encountered, since my father's copy, which he inherited from a relative, was on the bookshelf on the landing outside my childhood bedroom.

James Herriot, *If Only They Could Talk*, 1970

Jason Hill, *Wild Foods of Great Britain*, 1939: an early and inspirational book on foraging.

Barry Hines, *A Kestrel for a Knave*, 1968

W. G. Hoskins, *English Landscape*, 1977

W. H. Hudson, *Adventures Among Birds*, 1913

Richard Jefferies, *The Gamekeeper at Home*, 1878; *The Amateur Poacher*, 1879; *The Life of the Fields*, 1884

Rev. C. A. Johns, ed. J. A. Owen, *British Birds in Their Haunts*, 1938: every entry is nature literature.

Richard Lewington, *Pocket Guide to the Butterflies of Great Britain and Ireland*, 2003

Ronald Lockley, *The Private Life of the Rabbit*, 1964: the inspiration for Adams's *Watership Down*

Robert Macfarlane, *The Wild Places*, 2008

J. G. Millais, *The Natural History of British Game Birds*, 1909

Ian Moore, *Grass and Grasslands*, 1966: from the New Naturalist series.

Ernest Neal, *The Badger*, 1948: also from the New Naturalist series.

George Orwell, *Coming Up for Air*, 1939

Eric Parker, *Shooting Days*, 1918; *The Shooting Week-End Book*, n.d.

E. Pollard, M. D. Hooper and N. W. Moore, *Hedges*, 1974

Major Hesketh Prichard, *Sport in Wildest Britain*, 1921

Oliver Rackham, *The History of the Countryside*, 1986

Arthur Ransome, *Great Northern?*, 1947: In the last of the Swallows & Amazons series, the children believe the rare great northern diver is nesting on a Scottish loch. And that they must protect it.

Romany (G. Bramwell Evans), *A Romany in the Fields*, 1927; *Out with Romany by Meadow & Stream*, 1942

Siegfried Sassoon, *Memoirs of a Fox-Hunting Man*, 1928

Peter Scott, *The Eye of the Wind*, 1961: the man who founded the Wildfowl Trust commanded RN gunboats in World War Two, winning a Mention in Dispatches and a Distinguished Service Cross. His father was Scott of the Antarctic, who in his last frozen notes had asked his wife to 'make the boy interested in nature'.

John Seymour, *The Fat of the Land*, 1961

Henry Stephens, *The Book of the Farm*, 1844: the *vade mecum* of Victorian farming.

Paul Sterry, *Collins Complete Guide to British Wild Flowers*, 2006

David Streeter and Rosamond Richardson, *Discovering Hedgerows*, 1982

Thomas Traherne, *Centuries of Meditations*, 1908: Traherne might be described as 'The British St Francis of Assisi'.

Edward Thomas, *Collected Poems*, 1920

S. Vere Benson, *The Observer's Book of British Birds*, n.d.

Brian Vesey-Fitzgerald, *Game Birds*, 1946; *The Book of the Horse*, 1946

Paul Waring and Martin Townsend, *Concise Guide to the Moths of Great Britain and Ireland*, 2009

Gilbert White, *The Natural History of Selbourne*, 1789: the fountainhead of British nature-writing.

Raymond Williams, *The People of the Black Mountains*, Vols 1 & 2, 1989–90: the local boy made good. Williams hailed from the railway village of Pandy, at the bottom of the Black Mountains, to become a drama don at Cambridge and a Marxist philosopher. These two volumes of fiction can be self-consciously literary; nonetheless they articulate the history of the Black Mountains landscape in a way that really works.

Henry Williamson, *Tarka the Otter*, 1927

William Youatt, *Sheep: Their Breeds, Management and Diseases*, 1848

Music

J. S. Bach, *Sheep May Safely Graze*, 1713: the aria from the Hunt Cantata, in praise of the good shepherd.

Samuel Barber, *Adagio for Strings*, 1936

George Butterworth, *The Banks of Green Willow*, 1913: Butterworth was killed in action in 1916.

Hubert Parry, *Jerusalem*, 1916: yes, he was a relative; or at least claimed to be.

Henry Purcell, *When I Am Laid in Earth*, 1688: the aria from 'Dido and Aeneas'.

Supergrass, *Alright*, 1998: I should coco. Especially on a warm spring day when the grass is singing.

Ralph Vaughan Williams, *Fantasia on a Theme by Thomas Tallis*, 1910; *Folk Songs II: To The Green Meadow*, 1950

Thomas Tallis, *Spem in Alium*, c. 1570: 'Hope in any other'.

Acknowledgements

I would like to thank the Society of Authors for granting me a Foundation Award.

My thanks go to Susanna Wadeson and Patsy Irwin at Transworld, to Julian Alexander and Ben Clark at LAW, to my wife, to my children, and to all the not so dumb beasts of the field, wild and farmed, who tolerate me. To the flowers, grasses and trees too.

Lastly, I thank Faber & Faber and Viking for their permission to reproduce short extracts from, respectively, Ezra Pound's *The Classic Anthology Defined by Confucius* ('The Decade of Sheng Min') and John Stewart Collis's *The Worm Forgives the Plough*.

John Lewis-Stempel is a writer and farmer. His many previous books include *The Wild Life: A Year of Living on Wild Food*, *England: The Autobiography* and the bestselling *Six Weeks: The Short and Gallant Life of the British Officer in the First World War*. He lives on the borders of England and Wales with his wife and two children.

Also by John Lewis-Stempel:

England: The Autobiography
The Autobiography of the British Soldier
The Wild Life: A Year of Living on Wild Food
Six Weeks: The Short and Gallant Life of the British Officer in the First World War
Young Herriot
Foraging: The Essential Guide to Free Wild Food
The War Behind the Wire: The Life, Death and Glory of British PoWs, 1914–18